Navigating Local Sustainability in Food, Community, and Innovation

Miki Yoshizumi · Mariko Arata ·
Takaaki Miyaguchi
Editors

Navigating Local Sustainability in Food, Community, and Innovation

From Grassroots to Global

Springer

Editors
Miki Yoshizumi
College of Gastronomy Management
Ritsumeikan University
Kusatsu, Shiga, Japan

Mariko Arata
College of Gastronomy Management
Ritsumeikan University
Kusatsu, Shiga, Japan

Takaaki Miyaguchi
Faculty of Foreign Language Studies
Kansai University
Suita, Osaka, Japan

ISBN 978-981-96-8962-0 ISBN 978-981-96-8963-7 (eBook)
https://doi.org/10.1007/978-981-96-8963-7

This Springer imprint is published by the registered company Springer Nature Singapore Pte Ltd.
The registered company address is: 152 Beach Road, #21-01/04 Gateway East, Singapore 189721, Singapore

Preface I

As the Director of 'The Research Center for Gastronomic Arts and Sciences' (RCGAS), Ritsumeikan University, I feel extremely honoured to present this book *Navigating Local Sustainability in Food, Community, and Innovation: From Grass-roots to Global*. This book has been successful in materialising the long-term vision of the RCGAS. The RCGAS has always aspired to gain research-based knowledge related to food and establish and promote 'gastronomic sciences' as a comprehensive academic discipline, catering to diverse groups of ambitious students and research enthusiasts. In addition, we are actively engaged in dispersing and sharing this knowledge through a multitude of forums and ensuring the social implementation of our research results. The scope of our institution further incorporates activities, such as regional revitalisation, commercialisation, and internationalisation to encourage societal transformations and well-being. In January 2021, together with the Asia-Japan Research Institute Ritsumeikan University, the RCGAS organised a forum on food and agriculture which brought the international food and agriculture experts together on a single platform. This venture which instilled scientific temperament toward food sciences further paved the way for an online symposium which was held on 8th March 2021. The participants exchanged deep insights into the issues of food security and safety, regional resilience, and environmental impact of agriculture, which we face in the midst of global crises such as climate change and COVID-19. These discourses further acquainted us to the long-term solutions directed towards building a sustainable food system and eventually to set forth the views in the form of a book.

Thus, the original impetus for this book stemmed from this online forum held using a global scientific platform called *Meridian 180*, which can be accessed in multiple languages. The book includes various case studies focusing on sustainable solutions, such as food banks in Japan, local food systems in Indonesia, small-scale agriculture in Thailand, and environment friendly agriculture in Japan and other countries, each of which highlights the specific measures taken for the upliftment of the local community. These case studies will prove to be fundamental for the readers in understanding the implications of food-related challenges and to gain insights into the response mechanism needed to be adopted in such scenarios.

We hope this publication provides an enriching experience for readers worldwide, fostering new perspectives and practical knowledge on sustainable local food systems. The RCGAS is assured that this book will prove to be exemplary in the direction of addressing global food issues, ensuring equitable resource distribution, and fostering an economically sustainable future.

February 2025

Mariko Arata
Professor
Director of 'The Research Center for Gastronomic Arts and Sciences' (RCGAS)
College of Gastronomy Management
Ritsumeikan University
Kusatsu, Shiga, Japan

Preface II

Food is essential for the survival of all living beings. Human societies have been shaped around securing food since the ancient times. Societies have evolved by adapting to various natural limitations and changes while being built and shaped by specific ideologies and responses. Furthermore, as we progressed, advancements in agricultural technology and distribution systems have transformed the food production and supply chains, which, in turn, have further impacted the broader social structure.

Over the years, rising population, growing food demand, and altered lifestyles have caused globalisation to overtake local food production and supply systems. However, since 2020, the COVID-19 pandemic has restricted socioeconomic activities worldwide, disrupting global distribution systems. This has led to conflicts involving major agricultural nations that rely heavily on food imports for generating revenues. In addition, the import dependent nations are in a precarious position, sceptical of the future trajectory of their economy and food security. These events underscored the necessity of securing food at the regional level, rather than relying solely on globalised trades. At the same time, the Sustainable Development Goals (SDGs) have clearly reinforced this idea through a set of 17 objectives directing the individuals to bear the ownership of these goals while actively implementing on them. In particular, discussions on SDG implementation have emphasized the importance of local community initiatives.

This book outlines various ways the local communities can adopt to proactively address and implement the SDGs, including analysing the relevant case studies from around the world. It examines various community-led initiatives that directly and indirectly impact food systems, and are fundamental to human society. Through these case studies, the book establishes and illustrates the resilience of local food systems and their role in achieving the SDGs.

'Navigating Local Sustainability in Food, Community, and Innovation' was born out of a shared commitment to addressing these challenges through the context of local community set-up. The book's origins trace back to the research on local initiatives for sustainability that has been conducted at Kyoto University and partner institutions since 2016, where a diverse group of researchers shared their experience and expertise which unfolded a wide pool of global case studies. In April 2018, the establishment of the College of Gastronomy Management at Ritsumeikan University—the first academic institution in Japan dedicated to comprehensive and scientific research on food systems—provided the author with an opportunity to engage in interdisciplinary academic discussions on food-related issues. This eventually led to the organisation of the international symposium 'Beyond Food and Agriculture in the Post-COVID-19 Era' on March 8, 2021, in collaboration with colleagues from the College of Gastronomy Management.

A month-long online forum was conducted in partnership with Northwestern University through an international academic platform, *Meridian 180*. The online forum facilitated active discussions, which formed the foundation of the symposium. The event explored sustainable and insightful solutions to crises in the global food and agriculture sectors, particularly in light of the COVID-19 pandemic and climate change. A key focus of the discussions was on future courses of action that local communities could undertake in response to these challenges.

This book is structured based on the outcomes of these research activities. While expanding on the discussions from the symposium, the book delves into how demographic shifts and climate change impact food security and examines strategies to enhance community resilience and adaptive responses. By critically assessing and incorporating diverse perspectives, it aims to outline pathways that leverage local food systems to combat food insecurity and social disparities. To achieve these goals, the book presents case studies and research findings from a multitude of geographic regions, from rural Southeast Asia to Europe, offering both unique insights and universally applicable solutions.

Furthermore, the book suggests small-scale yet concrete initiatives that contribute to building a sustainable future and bridge the gap between the theoretical frameworks and their practical implementation. We hope this book helps readers gain a deeper understanding of the dynamics of local food systems and develop practical strategies that can be adapted to various circumstances. Each chapter underscores the necessity of an interdisciplinary approach that integrates environmental, social, and economic dimensions to ensure long-term sustainability.

I would like to express my deepest gratitude to all the experts who contributed through their knowledge and dedicated research to this book. Also special thanks to the Ritsumeikan University Institute of Asian-Japanese Research, whose support made this publication possible. The institute has served as a central hub for *Meridian 180* in Japan and has significantly contributed to the advancement of research on 'Scientific and Social Developments in Food and Agriculture', synthesizing valuable insights from various studies.

As you turn the pages of this book, I hope you will derive inspiration and gain practical perspective to support sustainable local food systems and contribute to the broader dissemination of these efforts. Through shared knowledge and collaborative initiatives, we can foster resilient, equitable, and sustainable food practices that are vital for the well-being of local communities and this planet as a whole.

February 2025

Miki Yoshizumi
Professor
College of Gastronomy Management
Ritsumeikan University
Kusatsu, Shiga, Japan

Takaaki Miyaguchi
Professor
Faculty of Foreign Language Studies
Kansai University
Suita, Osaka, Japan

Contents

Contributors

Siwaporn Chousorn Kyoto University ASEAN Center, Kyoto, Japan

Ayako Fujieda Faculty of Humanities, Kyoto Seika University, Kyoto, Japan

Thanh Tam Ho College of International Relations, Ritsumeikan University, Kyoto, Japan

Akiko Iizuka Center for International Exchange, Utsunomiya University, Utsunomiya, Japan

Thi Kinh Kieu The University of Danang, University of Science and Education, Danang, Vietnam

Hirohide Kobayashi Graduate School of Global Environmental Studies, Kyoto University, Kyoto, Japan

Motoki Kubo Department of Biotechnology, Faculty of Life Science, Ritsumeikan University, Shiga, Japan

Katalin Kuntner Independent Researcher, Budapest, Hungary

Van Anh Ngoc Le Faculty of Architecture, University of Sciences, Hue University, Hue, Vietnam

Aurora Lokita Resilients Cities Network, Singapore, Singapore

Takaaki Miyaguchi Faculty of Foreign Language Studies, Kansai University, Suita, Japan

Tung Ngoc Nguyen Faculty of Architecture, University of Sciences, Hue University, Hue, Vietnam

Nini Purwajati Resilients Cities Network, Singapore, Singapore

Holger Schaefer Graduate School of Global Environmental Studies, Kyoto University, Kyoto, Japan;
Department of Forest Soils, Forestry and Forest Products Research Institute, Tsukuba, Japan

Andarini Sertianti Independent Researcher, Kyoto, Japan

Koji Shimada College of Economics, Ritsumeikan University, Kyoto, Japan

Jane Singer Department of Global Studies, Kyoto University of Foreign Studies, Kyoto, Japan

Quoc Thinh Tran Department of Biotechnology, Faculty of Life Science, Ritsumeikan University, Shiga, Japan

Thuy Xuan Cam Vo The Center of Building Up Sustainability, Danang, Vietnam

Ary Wahyono Center for Society and Culture, National Research and Innovation Agency of Indonesia, Jakarta, Indonesia

Henny Warsilah Center for Society and Culture, National Research and Innovation Agency of Indonesia, Jakarta, Indonesia

Miki Yoshizumi College of Gastronomy Management, Ritsumeikan University, Kusatsu, Japan

List of Figures

Building Local Food Systems Through Non-rice Food Banks for Nutrition and Food Sustainability: An Indonesian Case Study

Self-Sustaining Community Resilience Through Local Asset Management: A Case Study of the Local Community Association 'Shinjyo Aigoukai' in Shinjyo Town of Tanabe City, Wakayama, Japan

Building Up Resilience for Small-Medium Enterprises in Traditional Food Villages: A Case Study of Central Vietnam

Potential and Challenges for Community Energy in the United Kingdom: A Case Study Approach

Issues and Challenges in Implementing Social Forestry Programs in Customary Communities: A Case Study of *Hutan Adat* in Indonesia

Sustainable Tourism in Japan: Kuriyama in Nikko City

Eco-Tourism Project in Hong Ha Commune: A Case Study of Sustainable Community Initiatives in Vietnam

Community Initiatives Drive Sustainable Resource Utilization Across Germany: Eco-tourism in Bad Hindelang, Environmental Education in Templin, and Power Generation in Schönau im Schwarzwald

Sustainable Agriculture for Small-Scale Farmers in Rural Thailand: The Practice of Farming for Self-sufficiency in Ban Na Isan

List of Tables

Eco-Tourism Project in Hong Ha Commune: A Case Study of Sustainable Community Initiatives in Vietnam

Community Initiatives Drive Sustainable Resource Utilization Across Germany: Eco-tourism in Bad Hindelang, Environmental Education in Templin, and Power Generation in Schönau im Schwarzwald

Soil Fertility of Paddy Fields in Japan and Organic Rice Cultivation Using SOFIX Analysis

Promotion Policies and Challenges in Sustainable Rice Agriculture in Shiga Prefecture, Japan

Global Insights and Local Dynamics

Introduction

Miki Yoshizumi and Takaaki Miyaguchi

Abstract This introductory chapter lays the conceptual groundwork for the volume Navigating Local Sustainability in Food, Community, and Innovation: From Grassroots to Global. In an era marked by the intensifying impacts of climate change, pandemics, and social inequality, it argues for the centrality of locally driven initiatives in achieving sustainable food systems and community resilience. Drawing on interdisciplinary theories—including food systems theory, resilience thinking, and social–ecological systems—the chapter explores how global challenges manifest at the local level and how community-based responses can offer scalable and context-sensitive solutions. It also introduces the book's four-part structure: (1) global insights and theoretical perspectives, (2) community practices enhancing food security and resilience, (3) sustainable innovations in energy, tourism, and disaster preparedness, and (4) agricultural transformations and policy responses. Together, these chapters highlight how local knowledge, collaboration, and innovation can serve as powerful tools for sustainability from the grassroots to the global scale.

Keywords Local sustainability · Food systems · Community resilience · Innovation · Social–ecological systems · Grassroots initiatives · SDGs · Interdisciplinary frameworks

1 The Imperative for Local Action in Food Systems

As articulated in various global frameworks, sustainable food systems aim to meet the present nutritional needs while safeguarding the ability of future generations to meet their own needs (Raza et al., 2020). This seemingly straightforward goal encompasses interconnected social, economic, and environmental challenges within

M. Yoshizumi (✉)
College of Gastronomy Management, Ritsumeikan University, Kusatsu, Japan
e-mail: miki-yo@fc.ritsumei.ac.jp

T. Miyaguchi
Faculty of Foreign Language Studies, Kansai University, Suita, Japan
e-mail: t_miyagu@kansai-u.ac.jp

M. Yoshizumi et al. (eds.), *Navigating Local Sustainability in Food, Community, and Innovation*, https://doi.org/10.1007/978-981-96-8963-7_1

food networks. Achieving sustainable food systems is crucial for attaining sustainable development and is included the Sustainable Development Goals (SDGs) (Varzakas & Smaoui, 2024). Climate change impacts agriculture, agrobiodiversity loss, food waste, malnutrition, and unequal access to food. The scope and complexity of these issues seem overwhelming on a global scale.

However, addressing these food system challenges has only been considered urgent recently. We are currently living in what several scientists refer to as the Anthropocene: a new geological epoch characterised by human domination of the planetary system (Malhi, 2017) and is marked by the dominant influence of anthropogenic activities, including industrial agriculture and global food trade, on Earth's climate and ecosystems (Steffen et al., 2011). The Anthropocene concept highlights that the impacts of anthropogenic activities on Earth are more encompassing than previously recognised, suggesting the end of the Holocene epoch (Castree, 2018). As the impacts of unsustainable food systems have become more apparent—from soil degradation and water scarcity to diet-related health crises and disappearance of traditional food cultures—there is a pressing need for transformation in food production, distribution, consumption patterns, and our relationship with the natural world.

However, translating this global imperative into concrete actions within food systems remains a formidable challenge. The sheer scale and interconnectedness of food sustainability issues may be too overwhelming and discouraging. International negotiations on food security and agricultural policy often progress slowly, hampered by competing national interests and the difficulty of coordinating actions across diverse food cultures and agricultural contexts, yet the window for averting the worst impacts of climate change on food production and other ecological crises continues to narrow.

In this context, the importance of local actions in building resilient communities—including but not limited to food systems—has become evident. While global coordination remains crucial, it is at the local level that both the impacts of unsustainable food practices and the potential for transformative solutions are most tangibly felt. Communities are at the front line of agricultural challenges, nutritional deficiencies, and food-related social strain. They experience the first-hand effects of changing weather patterns on crop yield, the adverse health effects of processed foods, and the social tensions that may arise from food insecurity and loss of culinary heritage.

Simultaneously, communities possess inherent knowledge of local food conditions, needs, and capabilities that inform effective food system interventions. They understand the nuances of local agroecosystems, intricacies of food-related social dynamics, and practical constraints and opportunities on the ground. This contextual knowledge is invaluable for developing effective and culturally appropriate solutions to issues concerning food availability and security.

Most importantly, communities can mobilise social networks, cultural resources, and the collective will to achieve shared food-related goals. The power of social capital—the bonds of trust, reciprocity, and cooperation within communities—should not be underestimated in the context of food systems (Glowacki-Dudka et al., 2013). Social capital plays a crucial role in sustainable local food networks and community development, influencing both the development of local food systems and how these systems contribute to building social capital (Glowacki-Dudka et al., 2013). When people come together with a common vision for a more sustainable and equitable food systems in the future, they can achieve remarkable results despite having limited resources. However, it is important to note that social capital and collective identity among local food movement organisations can both enhance and reduce collaboration, thereby impacting local food system sustainability (Bauermeister, 2016).

The ability of community-driven approaches to transform food systems has recently gained increasing recognition in both the academic literature and policy circles. The United Nations' SDG 2 (Zero Hunger) explicitly calls for 'local action' to achieve food security and promote sustainable agriculture. Hence, it represents a significant shift from top-down to bottom-up approaches to food and agricultural development, acknowledging that lasting changes must be rooted in and driven by local food communities.

Similarly, the implementation of various national food strategies relies heavily on sub-national and local efforts. Cities and regions worldwide are setting their own ambitious food sustainability targets, often exceeding national commitments. There is growing awareness that successfully addressing global food system challenges will require multilevel governance that empowers local actors while also facilitating coordination across scales of the food system (Janin et al., 2023).

This recognition of the importance of local action is reflected in emerging concepts like 'food sovereignty', which emphasises people's rights to define their own food and agriculture systems. Although global food frameworks and technologies are important, they must be adapted to and implemented in ways that respond to local food realities and values. However, the effective implementation of global policy agendas, such as the SDGs, at the local level requires a strong framework for multi-stakeholder engagement and coordination among different levels of government through both top-down and bottom-up approaches (Croese et al., 2021).

At the same time, communities are increasingly recognising the need to take action rather than waiting for government agencies to impose solutions to food-related issues. Faced with imminent threats to food security, nutrition, and culinary traditions, several localities have leveraged their resources and ingenuity to chart a more sustainable course for their food systems. These bottom-up initiatives often arise organically in response to local food needs and are driven by concerned citizens, community organisations, and local governments. However, research has identified tensions and conflicts between municipal and regional authorities, agricultural and food policies, and rural and urban territories, which can constrain the sustainability potential of multilevel agro-food policies (López-García et al., 2024).

However, grassroots food sustainability efforts can also be catalysed and supported by targeted interventions. Capacity-building programmes can help communities develop the skills and knowledge required to design and implement effective local food projects. Supportive policy frameworks can create conditions that enable local food innovation and experimentation. Knowledge-sharing platforms help successful local food models spread and adapt to different contexts. This approach aligns with recent research emphasising the importance of sharing community-derived good practices to support global networks of sustainable community food systems and facilitate wide-scale food system transformation (Blay-Palmer et al., 2015).

Engaging the youth in these processes is important. Increasing youth participation in creating sustainable food systems requires policies and practices that support food-system-related knowledge in two ways: democratising formal education systems and strengthening horizontal networks of grassroots research and innovation, including traditional, ecological, local, and community knowledge (Huambachano et al., 2022). This approach can enhance the sustainability and resilience of community food systems, while fostering knowledge co-creation and collective action to address global challenges (Blay-Palmer et al., 2015).

To address these challenges effectively, a more systematic analysis of food systems is required. This involves understanding the critical system interfaces, including intersections with other systems, interactions within the food system, and incentives for food system innovations (Ruben et al., 2018). Such an approach can aid in the development of programmes that enhance food system resilience, responsiveness, and inclusiveness.

The case studies explored in this book illustrate the diverse forms of community-based food sustainability initiatives. From food banks in rural Indonesia that are revitalising traditional crop varieties to community-supported agriculture schemes in Japan that are reconnecting consumers with local farmers, we see how local actors are reimagining the fundamental aspects of the food system to better serve both people and the planet. We examined community-based approaches to sustainability, including reducing food waste, preserving traditional knowledge, strengthening renewable energy networks, and fostering disaster resilience.

While each initiative is unique to its food context, common themes emerged regarding the importance of social capital, adaptive governance, and the integration of traditional and modern food knowledge systems. Strong community bonds and trust serve as critical enabling factors for collective action on food issues. We observed the value of flexible, learning-oriented approaches that allow experimentation with new food production and distribution models. We also note how several initiatives have been successful by integrating indigenous culinary wisdom with cutting-edge agricultural technologies and nutritional insights.

By examining these on-the-ground food efforts, we gained valuable insights into the mechanics of fostering sustainable local food systems. What factors enable some food initiatives to take root and scale while others struggle to gain traction? How can local food projects effectively interact with broader agricultural policy frameworks and food markets? What role can outside actors such as non-government organisations (NGOs), academic institutions, or government agencies play in supporting grassroots food sustainability without undermining local agencies or food sovereignty?

We also explored the challenges and limitations of community-driven food approaches. How can initiatives maintain their momentum and adapt over time, particularly in the face of changing food environments and dietary trends? How can power imbalances within food communities be addressed to ensure equitable participation and benefit sharing? What are the appropriate mechanisms for measuring and communicating the impact of local food sustainability initiatives?

Ultimately, the imperative for local action in sustainable food systems is not just efficacy, although community-driven approaches have demonstrated remarkable potential for creating context-appropriate food solutions. It is also about cultivating a sense of agency, connection, and shared purpose, which is essential for navigating the complex food challenges of the 21st century. By empowering communities to shape their own sustainable food futures, we tap into the wellspring of creativity, commitment, and culinary heritage, which may prove to be our most valuable resource for building more just and resilient local food systems.

2 Themes and Structure of the Book

This volume brings together case studies from Asia and Europe to provide a comparative perspective on community-led sustainability initiatives, exploring food systems alongside energy, tourism, disaster preparedness, and social resilience. While the specific focus of each project varies, several overarching themes emerged that help frame our analysis.

1. Local Food Systems and Food Security: Multiple chapters explore community efforts to build more resilient and sustainable local food systems. These include non-rice food banks in Indonesia and organic farming initiatives in Thailand. Food has emerged as a critical nexus in addressing the interconnected challenges of poverty, nutrition, cultural preservation, and environmental stewardship.
2. Community-Based Natural Resource Management: Several case studies examined how communities take greater control over the management of natural resources such as forests, fisheries, and water. These initiatives often combine traditional ecological knowledge with modern scientific approaches.
3. Alternative Energy Systems: Community-owned renewable energy projects showcase how local actors are driving the transition to cleaner, more decentralised power generation and distribution.
4. Sustainable Tourism: Multiple chapters describe community-based ecotourism initiatives that seek to generate economic benefits while preserving natural and cultural heritage sites.
5. Disaster Resilience: Some case studies focus on community-led efforts to enhance resilience to natural disasters through improved planning, education, and infrastructure.
6. Social Capital and Collective Action: A recurring theme across initiatives plays a critical role in social networks, trust, and cooperative governance structures in enabling effective collective action.

7. Integration of Traditional and Modern Knowledge: Several of the featured projects showcase the creative integration of indigenous and traditional knowledge with modern scientific and technological approaches.
8. Multi-Stakeholder Partnerships: Successful initiatives often involve productive partnerships between community groups, local governments, NGOs, and sometimes private-sector actors.

The book is structured to progress from focused sectoral initiatives to broader, integrated approaches to community sustainability. It begins with several chapters that examine local food system initiatives that provide a tangible entry point for exploring the core concepts of community resource management and resilience. Subsequent sections examine community-based approaches to energy, tourism, disaster preparedness, and holistic rural development.

Throughout this book, we aimed to balance the rich description of on-the-ground realities by analysing the broader patterns and lessons learned. Each case study provides a historical and cultural context, details of project implementation and governance, and an assessment of the outcomes and challenges faced. A comparative analysis across cases allowed us to identify common and context-specific success factors.

The conclusion synthesises the key findings and explores their implications for policy, practice, and further research on community-driven sustainability. It reflects on how local initiatives can be more effectively supported and scaled while retaining the participatory, context-sensitive character that is at the core of their success.

As the global community grapples with accelerating environmental changes and social challenges, the role of local action in advancing sustainability is likely to grow in importance. The stories and lessons captured in this volume provide inspiration and practical insights into this vital work. By amplifying these diverse voices from the front lines of community-based sustainability, we hope to catalyse further grassroots innovation while enriching broader dialogues on pathways to a more sustainable and equitable world.

References

Bauermeister, M. R. (2016). Social capital and collective identity in the local food movement. https://doi.org/10.1080/14735903.2015.1042189

Blay-Palmer, A., Sonnino, R., & Custot, J. (2015). A food politics of the possible? Growing sustainable food systems through networks of knowledge. *Agriculture and Human Values*. https://doi.org/10.1007/s10460-015-9592-0

Castree, N. (2018). Anthropocene and planetary boundaries. https://doi.org/10.1002/9781118786352.WBIEG0027

Croese, S., Oloko, M., Simon, D., & Valencia, S. C. (2021). Bringing the global to the local: The challenges of multi-level governance for global policy implementation in Africa. *International Journal of Urban Sustainable Development*. https://doi.org/10.1080/19463138.2021.1958335

Glowacki-Dudka, M., Murray, J., & Isaacs, K. P. (2013). Examining social capital within a local food system. https://doi.org/10.1093/CDJ/BSS007

Huambachano, M., Arulingam, I., Bowness, E., Korzenszky, A., Mungai, C., Termine, P., & Wittman, H. (2022). Knowledge networks to support youth engagement in sustainable food systems. *Frontiers in Sustainable Food Systems.* https://doi.org/10.3389/fsufs.2022.867344

Janin, P., Nzossié, E. F., & Racaud, S. (2023). Governance challenges for sustainable food systems: The return of politics and territories. *Current Opinion in Environmental Sustainability.* https://doi.org/10.1016/j.cosust.2023.101382

López-García, D., Cruz-Maceín, J. L., & DiPaula, M. (2024). Agri versus food? Perceptions of local policymakers on agro-food policies from a multilevel approach. *Frontiers in Sustainable Food Systems.* https://doi.org/10.3389/fsufs.2024.1399746

Malhi, Y. (2017). The concept of the anthropocene. https://doi.org/10.1146/ANNUREV-ENVIRON-102016-060854

Raza, A., Fox, E. L., Morris, S. S., Kupka, R., Timmer, A., Dalmiya, N., & Fanzo, J. (2020). Conceptual framework of food systems for children and adolescents. *Global Food Security.* https://doi.org/10.1016/j.gfs.2020.100436

Ruben, R., Verhagen, J., & Plaisier, C. (2018). The challenge of food systems research: What difference does it make? *Sustainability.* https://doi.org/10.3390/SU11010171

Steffen, W., Persson, Å., Deutsch, L., Zalasiewicz, J., Williams, M., Richardson, K., & Svedin, U. (2011). The anthropocene: From global change to planetary stewardship. *Ambio.* https://doi.org/10.1007/s13280-011-0185-x

Varzakas, T., & Smaoui, S. (2024). Global food security and sustainability issues: The road to 2030 from nutrition and sustainable healthy diets to food systems change. *Foods.* https://doi.org/10.3390/foods13020306

A Comprehensive Review of Local Food Systems and Their Resilience

Miki Yoshizumi and Takaaki Miyaguchi

Abstract This chapter offers a critical synthesis of the current scholarly and policy discourse on the resilience of local food systems in the face of global disruptions such as climate change, pandemics, and economic instability. It introduces an integrated theoretical framework grounded in food systems theory, resilience thinking, and social-ecological systems analysis to assess the structural and functional characteristics that support adaptability and sustainability. Through the review, four essential elements of resilient food systems are identified: production diversity, localized distribution networks, community participation, and the incorporation of traditional ecological knowledge. In addition to outlining these enabling factors, the chapter examines cross-cutting challenges—including environmental uncertainty, policy fragmentation, market limitations, and demographic transitions—that threaten the viability of localized food networks. By linking conceptual insights to actionable strategies, this chapter serves as a foundation for understanding and contextualizing the applied case studies presented in the chapters that follow.

Keywords Food system resilience · Localized supply chains · Adaptive capacity · Community-based food networks · Traditional ecological knowledge · Food governance · Sustainability transitions · Social-ecological systems · Food security policy

---

M. Yoshizumi (✉)
College of Gastronomy Management, Ritsumeikan University, Kusatsu, Japan
e-mail: miki-yo@fc.ritsumei.ac.jp

T. Miyaguchi
Faculty of Foreign Language Studies, Kansai University, Suita, Japan
e-mail: t_miyagu@kansai-u.ac.jp

M. Yoshizumi et al. (eds.), *Navigating Local Sustainability in Food, Community, and Innovation*, https://doi.org/10.1007/978-981-96-8963-7_2

1 Introduction

The United Nations (UN) Food Systems Summit, which convened at the UN General Assembly, New York, in September 2021, underscored the urgent need to transform the global food system to achieve the Sustainable Development Goals. The COVID-19 pandemic has been highlighted as a pivotal event, revealing unprecedented disruptions in food supply and demand owing to large-scale social and economic disturbance. These disruptions intensified existing inequalities, as many lost jobs and incomes disproportionately affected vulnerable populations in both urban and rural areas. Consequently, vulnerabilities in centralised food supply chains have been exposed, which has emphasised the need for more localised, resilient food systems that can mitigate the risk of supply disruptions. Local food systems have increasingly emerged as critical elements in agricultural and food policy discussions, offering potential pathways for strengthening global food security, promoting environmental sustainability, and enhancing community resilience (Béné, 2020). Thus, regional or local level food systems can enable communities to have greater control over their food security in times of crisis.

As highlighted by the UN Food Systems Summit, the present moment must be seized to build stronger, more resilient food systems that can withstand future challenges (UN Food Systems Summit, 2021). The pandemic has offered a rare opportunity to reimagine food systems in manners that enhance their efficiency, resilience, health, sustainability, and equity. The continued global crisis serves as a lesson and call for action to policymakers to drive structural change across food systems, adapting to multiple shocks, including health crises, climate change, economic instability, and conflict, through cross-sectoral collaboration.

This chapter presents a comprehensive review of prior research on local food systems, with particular attention paid to their resilience in addressing climate change, natural disasters, demographic shifts, and the ongoing threat of infectious disease outbreaks. This introduction explores the fundamental importance of local food systems and the concept of food resilience, setting the stage for a thorough examination of the strategies and practices that can enhance resilience across local food networks.

1.1 Importance of Local Food Systems

Local food systems are defined by their short supply chains, emphasis on sustainability, and focus on locality and proximity (Souza et al., 2020). Despite such characteristics, the concept of 'local food' is inherently complex and multifaceted, lacking a universally accepted definition (Martinez et al., 2010). These systems encompass not only the production and consumption of food within a specific geographic area but also the social, economic, and environmental relationships that develop around food (Feagan, 2007). Chicoine et al. (2022) proposed that local food can be defined through nine dimensions of perceived proximity: geography, process, price, identity, relation, function, culture, access, and experience.

Local food systems can generate economic gains for producers, consumers, and communities by offering farmers opportunities to diversify production, add value to products, and ensure a more stable income (Souza et al., 2020). They also support local economies by maintaining money circulation within the communities. For small-scale farmers, local food systems provide an opportunity to differentiate products and access niche markets (Martinez et al., 2010). Furthermore, local food systems contribute to community well-being by fostering social connections, preserving cultural traditions, and promoting a sense of place (Ostrom, 2006). They can also enhance food security by increasing access to fresh and culturally appropriate food. The social and cultural aspects of local food systems are often tied to concepts of identity, tradition, and community cohesion (Feagan, 2007).

Although local food systems are often associated with reduced food miles and environmental benefits, research suggests that the relationship between locality and sustainability is complex (Stein & Santini, 2022). Factors such as production methods, consumer dietary choices, and resource management play significant roles in determining overall environmental impacts. The sustainability of local food systems is influenced by a range of factors beyond transportation, such as land use, agricultural production methods, resource efficiency, and consumer behaviour (Gregory et al., 2005).

As global food systems face increasing pressures from climate change and population growth, local food systems offer potential pathways for enhancing resilience and adaptability (Gregory et al., 2005). However, the ability of local food systems to effectively address food security challenges hinges on policy support, robust community engagement, and their strategic integration into global food networks. This integration requires a nuanced approach that recognises the complex interactions between location-based food practices and large-scale governance structures (Sonnino et al., 2016).

1.2 Concept of Food Resilience

Although resilience and sustainability are related concepts, they are not identical. Resilience focuses on the ability to withstand and recover from shocks, whereas sustainability emphasises long-term viability and environmental stewardship (Stein & Santini, 2022). Both concepts are crucial for the development of robust and enduring food systems.

The concept of food resilience is rooted in ecological and social resilience theories and adapts these ideas to the context of food systems (Béné, 2020). Food resilience refers to the capacity of a food system to withstand and recover from various shocks and stressors while maintaining its essential functions (Gregory et al., 2005). Resilient food systems are characterised by diversity, redundancy, and adaptability (Béné, 2020). These systems maintain food production, distribution, and access in the face of various challenges, including climate change, economic shocks, and social disruptions (Gregory et al., 2005).

Measuring food system resilience remains a challenge owing to the complexity of food systems and their diverse range of potential shocks and stressors (Béné, 2020). Although various frameworks and indicators have been proposed, no universally accepted method exists to quantify the resilience of food systems. Additionally, the interconnectedness of global food systems implies that local resilience efforts must be considered in a broader context (Sonnino et al., 2016).

1.3 Scope and Objectives

This review focuses on local food systems and food resilience in developed countries, with particular emphasis on North America and Europe. The temporal scope covers literature from the past two decades, reflecting the recent surge in interest and research in this field. This review aims to provide a comprehensive synthesis of the current knowledge on local food systems and food resilience by identifying key trends, research gaps, and areas for future research. The findings of this review will contribute to the ongoing debates on the role of local food systems in addressing global food security challenges and building resilient communities.

2 Theoretical Framework

The integration of food systems theory, resilience theory, and social-ecological systems (SES) approaches offers a comprehensive framework for understanding and enhancing local food system dynamics. This section explores these theoretical underpinnings and their applications in building resilient, sustainable, and equitable local food systems.

2.1 Food Systems Theory

Food systems theory has evolved from linear models that focus primarily on production and consumption to more complex and interconnected frameworks that consider the entire food chain and its broader societal and environmental contexts (James & Friel, 2015). This evolution reflects the growing recognition of the multifaceted nature of food systems and their impacts on health, economics, and ecology. Food systems theory has increasingly recognised the role of power dynamics in shaping food production, distribution, and access. This includes examining how various actors, from large corporations to small-scale farmers, influence food system outcomes, and how policies and economic structures create inequities within the system (Doherty et al., 2019).

Modern food system theory encompasses various interconnected components, including production, distribution, consumption, and waste management. It also considers the social, economic, and environmental factors influencing these components (Hodbod & Eakin, 2015). Such an approach allows a more comprehensive understanding of the challenges and opportunities within food systems. Although traditional linear models emphasise a straightforward path from production to consumption, circular models have gained prominence because of their focus on sustainability and resource efficiency. Circular models emphasise the importance of reducing waste, recycling nutrients, and creating closed-loop systems that minimise environmental impacts (Wood et al., 2023).

The application of food systems theory to local contexts involves adapting broader principles to specific geographic, cultural, and economic conditions. This localisation allows for more targeted interventions and policies that address unique challenges and opportunities in a given area (Ballamingie et al., 2020).

2.2 *Resilience Theory*

Resilience theory emerged as a means to understand how systems respond to disturbances and maintain their essential functions (Berkes & Ross, 2013). It originated in ecology, but has since been applied to various fields, including food systems. In the context of food systems, resilience theory provides a framework for analysing and enhancing the ability of these systems to withstand and adapt to shocks and stresses. The core principles of resilience are diversity, redundancy, connectivity, and adaptive capacity. These principles emphasise the importance of having multiple pathways and resources within a system as well as the ability to learn and adapt in the face of change (Béné, 2020). In food systems, this translates into practices such as crop diversification, social network building, and flexible supply chain development.

The adaptive cycle and panarchy concepts in the resilience theory describe how systems undergo cycles of growth, conservation, release, and reorganisation across multiple scales. This understanding is crucial for food systems as it helps explain how local food networks can adapt to changes at both the local and global levels (Bullock et al., 2017). Furthermore, resilience theory distinguishes between the concepts of resilience and vulnerability, highlighting that reducing vulnerability does not necessarily increase resilience. This nuance is important in food system planning, as it suggests that strategies should focus not only on protecting against specific threats, but also on building the overall system capacity to respond to unforeseen challenges (Milestad et al., 2010).

Despite the valuable insights it offers, resilience theory has been criticised for potentially reinforcing existing power structures and overlooking social justice issues. In the context of food systems, ensuring that resilience-building efforts address issues of equity and access rather than simply maintaining the status quo is crucial (Jones et al., 2022).

2.3 SES Approach

The SES approach emerged from the recognition that human and natural systems are inextricably linked and must be studied as integrated wholes. This perspective is particularly relevant to food systems, which are deeply embedded in social and ecological contexts (Helenius et al., 2007). SES frameworks typically include both social (e.g., institutions, economic systems, and cultural practices) and ecological (e.g., biodiversity, ecosystem services, and natural resources) components. This involves examining the effect of agricultural practices on local ecosystems and influence of social norms on food consumption patterns (Duncan et al., 2018).

Understanding feedback loops and thresholds is crucial for SES analysis. In food systems, this might involve studying how changes in land use affect soil quality, which in turn affects agricultural productivity and food security. Identifying critical thresholds can help prevent system collapse and guide sustainable management practices (Skog et al., 2018). The SES approach offers valuable tools for enhancing the resilience of local food systems. The consideration of the complex interactions between social and ecological factors enable holistic strategies for building adaptive capacity and addressing vulnerabilities in local food networks (Stone et al., 2024).

3 Key Components of Resilient Local Food Systems

3.1 Production Diversity

Production diversity is a crucial component of a resilient local food system as increasing agricultural diversity can lead to greater resilience in food systems (Macfall et al., 2015). This diversity can be considered at multiple levels, including genetic diversity in crop varieties, species diversity through mixed cropping, and broad ecosystem diversity across farms and food systems (Johns et al., 2013). Diverse agricultural practices, such as biointensive cultivation, can increase the resilience and sustainability of food systems by improving water use efficiency, yield, and nutrient retention, while reducing pressure from pests and pathogens (Macfall et al., 2015). Integrating livestock production with crop production can further enhance system diversity and resilience. This integration can provide additional income streams and contribute to nutrient cycling in farming systems (Johns et al., 2013).

Diversified production systems offer economic benefits to farmers by reducing risk and providing multiple income streams. High-value food crops within domestic and global value chains can be entry points for smallholder participation as contributors to and beneficiaries of development (Johns et al., 2013). However, despite these benefits, maintaining production diversity remains challenging. Pressures from market demand for specific crops and the complexity of managing diverse systems can pose obstacles for farmers (Macfall et al., 2015).

3.2 *Short Supply Chains*

Short food supply chains (SFSCs) aim to reconnect producers to local consumers and markets. They are characterised by reduced distance between producers and consumers, both geographically and in terms of intermediaries (Jarzębowski et al., 2020). SFSCs often employ direct marketing strategies such as farmers' markets, farm shops, and community-supported agriculture schemes to allow producers to capture more value from their products and build relationships with consumers (Jarzębowski et al., 2020). Notably, food hubs and cooperative distribution models can play a crucial role in SFSCs by aggregating products from multiple small producers and facilitating efficient distribution in local markets (Michel-Villarreal, 2022).

SFSCs can have positive environmental impacts, including reduced food miles, less energy use, lower carbon footprint, and decreased food waste (Jarzębowski et al., 2020). However, it should be noted that local food is not automatically more sustainable, as many factors beyond transportation affect the environment (Stein & Santini, 2022). Although SFSCs offer many benefits, they also face challenges such as logistical difficulties, limited product range, and seasonality issues. Additionally, the economic viability of these systems is a concern, particularly for small-scale producers (Michel-Villarreal, 2022).

3.3 *Community Engagement*

Community engagement is essential to building resilient local food systems. This can take various forms, including participation in local food production, distribution, and decision-making processes (Feenstra, 1997). For instance, community-supported agriculture (CSA) models are a form of community engagement in which consumers share the risks and rewards of food production with farmers. These models can strengthen connections between producers and consumers and increase trust and sense of community (Jarzębowski et al., 2020).

Local food policy councils and participatory governance structures can empower communities in shaping their food systems. These bodies can facilitate collaboration among diverse stakeholders and inform policy decisions (Feenstra, 1997). Education and awareness programs also play crucial roles in community engagement. These initiatives increase our understanding of local food systems, sustainable practices, and the importance of supporting local producers (Feenstra, 1997).

Notable barriers to community engagement include lack of time, resources, or knowledge. Addressing these barriers requires targeted strategies and support mechanisms to facilitate participation (O'Connor et al., 2024).

3.4 Traditional Knowledge Integration

Traditional and indigenous knowledge, which encompasses a deep understanding of the local ecosystems, sustainable farming practices, and food preservation techniques (Johns et al., 2013), can contribute significantly to the resilience of local food systems. Traditional ecological knowledge often emphasises biodiversity, ecosystem health, and long-term sustainability (Bohensky & Maru, 2011). Such knowledge can inform sustainable agricultural practices that are well-adapted to local conditions.

The effective integration of traditional and scientific knowledge can enhance the resilience of food systems. This integration requires new frames for collaboration, greater recognition of social contexts, expanded modes of knowledge evaluation, and involvement of intercultural 'knowledge bridgers' (Bohensky & Maru, 2011). Furthermore, the preservation of traditional food practices can contribute to cultural resilience and biodiversity conservation as these practices often maintain crop diversity and support sustainable land management techniques (Johns et al., 2013).

Challenges in preserving and applying traditional knowledge include the loss of cultural transmission, lack of recognition in formal policy frameworks, and difficulties in adapting traditional practices to changing environmental and social conditions (Bohensky & Maru, 2011).

4 Challenges to Local Food System Resilience

Local food systems face numerous challenges in testing their resilience to changing environmental, economic, social, and policy conditions. This section examines four key areas of concern: climate change impacts, economic pressures, policy barriers, and demographic shifts.

4.1 Climate Change Impacts

Climate change poses significant threats to local food systems by affecting agricultural productivity and food security. Shifts in temperature and precipitation patterns can significantly affect crop yield and livestock production. Kelly and Schulschenk (2011) note that the effects of climate change and cheap oil consumption exacerbated the instability of the global food system. This underscores the vulnerability of the local food systems to climate change and resource constraints. James and Friel (2015) emphasised the importance of building equity and resilience in local food systems in response to ongoing global climate change and acute crises such as the COVID-19 pandemic, highlighting the need for adaptive strategies to address these challenges.

The increased frequency and intensity of extreme weather events such as floods, droughts, and storms can disrupt food production, distribution, and access. Stone et al. (2024) found that local producers demonstrated notable resilience during the COVID-19 pandemic, which they categorised as human capital. This suggests that building adaptive capacity within local food systems can help mitigate the impact of unexpected shocks.

Climate change has also exacerbated water scarcity, posing challenges to irrigation and sustainable water management in agriculture. James and Friel (2015) suggested that agriculture should be managed as an ecosystem that requires certain water flows and provides essential ecosystem services. This approach could help address water-related challenges in local food systems.

4.2 Economic Pressures

Economic factors play crucial roles in the resilience of local food systems by influencing their ability to compete with global markets and maintain financial sustainability. Local food systems often struggle to compete with the economies of scale achieved by global food chains. Duncan et al. (2018) suggested that policy support for regional food networks plays a role in economic development and resilience by diversifying rural economies. This indicates the potential for policy interventions to strengthen local food economies in the face of global competition.

Small-scale producers often face financial challenges that limit their ability to invest in sustainable practices and technologies. Kelly and Schulschenk (2011) identified the dominance of large-scale commercial agriculture as a key vulnerability in the food systems they studied. This highlights the need for strategies that support smaller producers in local food systems. However, implementing sustainable practices can be costly for local food producers, requiring careful consideration of the long-term benefits and short-term expenses. Carducci et al. (2023) argued the necessity of new science, policy reorientation, and investments in environmental, technological, and sociocultural transformation to address these challenges. Consequently, a comprehensive approach to evaluate and support sustainable practices in local food systems is necessary.

4.3 Policy Barriers

Policy and regulatory frameworks significantly affect the development and resilience of food systems. Access to suitable farmland is crucial for local food production. Kelly and Schulschenk (2011) identified the dominance of large-scale commercial agriculture as a key vulnerability, suggesting that land use policies may favour larger operations over small-scale local producers.

Food safety regulations can pose challenges to small-scale producers. Stone et al. (2024) found that local food systems have significant weaknesses in terms of equity and resilience, particularly in terms of cultural, social, natural, and political capital. This suggests that policy frameworks may not always adequately support small-scale local producers.

Agricultural subsidies can also influence the competitiveness of local food systems. Duncan et al. (2018) argued that policy support for regional food networks may play a role in economic development and resilience by diversifying rural economies, indicating the potential of targeted subsidies in strengthening local food networks.

4.4 Demographic Shifts

Changes in population dynamics and consumer preferences significantly impact local food system resilience. Increased urbanisation affects food production and distribution patterns. Berkhout et al. (2023) argued that strengthening urban-rural connections is essential for building resilient food systems that can withstand various shocks and stressors, including those induced by climate change. By fostering agency, buffering capacity, connectivity, and diversification within food systems, urban-rural linkages enhance the capacity of communities to ensure access to sufficient, nutritious, and sustainably produced food. Their analysis underscores the role of secondary cities as central nodes, digital infrastructure as facilitators, and informal actors such as street food vendors in creating adaptive and inclusive food systems that benefit individuals, communities, and ecosystems.

Consumer preferences influence the demand for local and sustainable food products. Stone et al. (2024) suggested that future improvements should involve changes in infrastructure, markets, and cultural values to enhance the equity and resilience of local urban food systems. Thus, local food systems must adapt to changing consumer trends while maintaining their core values.

In summary, addressing the challenges of local food resilience requires a multifaceted approach that considers climate change impacts, economic pressures, policy barriers, and demographic shifts. Stakeholders can build resilient and sustainable local food systems by understanding and addressing these challenges.

5 Conclusion

5.1 *Summary of Key Findings*

This comprehensive review of the research on local food system resilience reveals the multifaceted nature of building sustainable and resilient local food networks. The key findings underscore the importance of integrating food systems theory, resilience theory, and SES approaches to develop a holistic understanding of local food dynamics. This review highlights that resilient local food systems are characterised by production diversity, short supply chains, strong community engagement, and traditional knowledge integration (Béné, 2020; Johns et al., 2013).

Specifically, production diversity is a crucial component that enhances system resilience through improved water-use efficiency, yield, and nutrient retention while simultaneously reducing vulnerability to pests and pathogens. SFSCs have demonstrated the potential to reduce environmental impacts and strengthen producer–consumer relationships, although their implementation faces logistical and economic challenges. Furthermore, community engagement, manifested through models such as community-supported agriculture and food policy councils, plays a vital role in building system resilience by fostering trust and participatory decision making. The integration of traditional ecological knowledge with scientific approaches also offers a promising pathway for developing sustainable agricultural practices that can be adapted to local conditions.

5.2 *Implications for Policy and Practice*

The findings of this review have significant implications for local food system policies and practices. Policymakers should consider developing frameworks that support production diversity and short supply chains, while addressing the economic pressures faced by small-scale producers. This involves reassessing land use policies, food safety regulations, and subsidy structures to create a more favourable environment for local food systems.

Practitioners of local food systems should focus on building adaptive capacity to address challenges such as climate change impacts and changing consumer preferences. This involves implementing diverse production strategies, strengthening community networks and integrating traditional knowledge into modern agricultural practices. Furthermore, innovative approaches must overcome logistical challenges in short supply chains and enhance the economic viability of local food initiatives.

5.3 Call for Integrated Research and Action

The complexities of local food system resilience require an integrated approach to research and action. Future research should aim to bridge the gaps between food systems theory, resilience thinking, and SES approaches. Interdisciplinary studies should examine the interactions between the different components of local food systems and their impact on overall resilience. Moreover, action-oriented research that engages directly with local communities and food system stakeholders is crucial for the development of practical, context-specific solutions. This could involve participatory research methodologies and the co-creation of knowledge among researchers, practitioners, and community members. Finally, long-term studies that track the evolution of local food systems over time, particularly in response to global challenges such as climate change and economic instability, should be urgently conducted. Such research could provide valuable insights into the adaptive capacities of local food systems and more effective strategies for building resilience in diverse contexts.

References

Ballamingie, P., Blay-Palmer, A., Knezevic, I., Lacerda, A. E., Nimmo, E. R., Stahlbrand, L. M., & Ayalon, R. (2020). Integrating a food systems lens into discussions of urban resilience: A policy analysis. *Journal of Agriculture, Food Systems, and Community Development, 9*(3), 159–176.

Béné, C. (2020). Resilience of local food systems and links to food security—A review of some important concepts in the context of COVID-19 and other shocks. *Food Security, 12*, 805–822.

Berkes, F., & Ross, H. (2013). Community resilience: Toward an integrated approach. *Society Natural Resources, 26*(1), 5–20.

Berkhout, E., Sovová, L., & Sonneveld, A. (2023). The role of urban–rural connections in building food system resilience. *Sustainability, 15*(3), 1818.

Bohensky, E. L., & Maru, Y. (2011). Indigenous knowledge, science, and resilience: What have we learned from a decade of international literature on "integration"? *Ecology and Society, 16*(4), 6.

Bullock, J. M., Dhanjal-Adams, K. L., Milne, A., Oliver, T. H., Todman, L. C., Whitmore, A. P., & Pywell, R. F. (2017). Resilience and food security: Rethinking an ecological concept. *Journal of Ecology, 105*(4), 880–884.

Carducci, B., Jägermeyr, J., Ruane, A., & Fanzo, J. (2023). Rising to the challenge: Producing and sustaining a nutrient-dense and climate-resilient food basket for all. *One Earth, 6*(11), 1443–1446.

Chicoine, M., Rodier, F., & Durif, F. (2022). Local food: A constellation of perceived proximity. *British Food Journal, 124*(13), 163–181.

Doherty, B., Ensor, J., Heron, T., & Prado, P. (2019). Food systems resilience: Towards an interdisciplinary research agenda. *Emerald Open Research 1*(10).

Duncan, S. L., Brekken, C. A., Lurie, S., Fiegener, R., Sherry, S., & Liang, C. (2018). Can regional food networks and entrepreneurial strategies enhance food system resilience? *Choices, 33*(2), 1–10.

Feagan, R. (2007). The place of food: Mapping out the local in local food systems. *Progress in Human Geography, 31*(1), 23–42.

Feenstra, G. W. (1997). Local food systems and sustainable communities. *American Journal of Alternative Agriculture, 12*(1), 28–36.

Gregory, P. J., Ingram, J. S., & Brklacich, M. (2005). Climate change and food security. *Philosophical Transactions of the Royal Society b: Biological Sciences, 360*(1463), 2139–2148.

Helenius, J., Aro-Heinilä, E., Hietala, R., Mikkola, M., Risku-Norja, H., Seppänen, L., & Vihma, A. (2007). Systems frame for multidisciplinary study on sustainability of localising food. *Progress in Industrial Ecology, an International Journal, 4*(5), 328–347.

Hodbod, J., & Eakin, H. (2015). Adapting a social-ecological resilience framework for food systems. *Journal of Environmental Studies and Sciences, 5*(3), 474–484.

James, S. W., & Friel, S. (2015). An integrated approach to identifying and characterising resilient urban food systems to promote population health in a changing climate. *Public Health Nutrition, 18*(13), 2498–2508.

Jarzębowski, S., Bourlakis, M., & Bezat-Jarzębowska, A. (2020). Short food supply chains (SFSC) as local and sustainable systems. *Sustainability, 12*(11), 4715.

Johns, T., Powell, B., Maundu, P., & Eyzaguirre, P. B. (2013). Agricultural biodiversity as a link between traditional food systems and contemporary development, social integrity and ecological health. *Journal of the Science of Food and Agriculture, 93*(14), 3433–3442.

Jones, S. D., Krzywoszynska, A., & Maye, D. (2022). Resilience and transformation: Lessons from the UK local food sector in the COVID-19 pandemic. *The Geographical Journal, 188*(1), 72–86.

Kelly, C., & Schulschenk, J. (2011). Assessing the vulnerability of Stellenbosch's food system and possibilities for a local food economy. *Development Southern Africa, 28*(4), 563–578.

Macfall, J., Lelekacs, J., LeVasseur, T., Moore, S., & Walker, J. (2015). Toward resilient food systems through increased agricultural diversity and local sourcing in the Carolinas. *Journal of Environmental Studies and Sciences, 5*(4), 608–622.

Martinez, S., Hand, M., Da Pra, M., Pollack, S., Ralston, K., Smith, T., & Newman, C. (2010). Local food systems: Concepts, impacts, and issues. Economic Research Report No. 97, U.S. Department of Agriculture, Economic Research Service. Available at: https://www.ers.usda.gov.

Michel-Villarreal, R. (2022). Towards sustainable and resilient short food supply chains: A focus on sustainability practices and resilience capabilities using case study. *British Food Journal, 124*(13), 210–226.

Milestad, R., Westberg, L., Geber, U., & Björklund, J. (2010). Enhancing adaptive capacity in food systems: Learning at farmers markets in Sweden. *Ecology and Society, 15*(3).

O'Connor, G., Reis, K., Desha, C., & Burkett, I. (2024). Valuing farmers in transitions to more sustainable food systems: A systematic literature review of local food producers experiences and contributions in short food supply chains. *Agriculture and Human Values* 1–28.

Ostrom, M. (2006). Everyday meanings of local food: Views from home and field. *Community Development, 37*(1), 65–78.

Skog, K. L., Eriksen, S., Brekken, C. A., & Francis, C. (2018). Building resilience in social-ecological food systems in Vermont. *Sustainability, 10*(12), 4813.

Sonnino, R., Marsden, T., & Moragues-Faus, A. (2016). Relationalities and convergences in food security narratives: Towards a place-based approach. *Transactions of the Institute of British Geographers, 41*(4), 477–489.

Souza, A. B., Fornazier, A., & Delgrossi, M. E. (2020). Local food systems: Potential for new market connections for family farming. *Ambiente & Sociedade, 23*, Article e02482.

Stein, A. J., & Santini, F. (2022). The sustainability of local food: A review for policy-makers. *Review of Agricultural, Food and Environmental Studies, 103*, 77–89.

Stone, T. F., Huckins, E. L., Hornbuckle, E. C., Thompson, J. R., & Dentzman, K. E. (2024). Equity and resilience in local urban food systems: A case study. *Agriculture and Human Values, 41*, 1239–1256.

UN Food Systems Summit (2021). COVID-19 pandemic offers rare chance for food systems transformation. Available at: https://www.un.org/en/food-systems-summit/news/covid-19-pandemic-offers-rare-chance-food-systems-transformation.

Wood, A., Queiroz, C., Deutsch, L., González-Mon, B., Jonell, M., Pereira, L., & Wassénius, E. (2023). Reframing the local–global food systems debate through a resilience lens. *Nature Food, 4*(2), 89–91.

Community Resilience and Food Security

Building Local Food Systems Through Non-rice Food Banks for Nutrition and Food Sustainability: An Indonesian Case Study

Henny Warsilah and Ary Wahyono

Abstract The Food and Agriculture Organization (FAO) has determined that rice is a key food worldwide, consumed by approximately three billion people daily. Rice is a staple food for 60% of the world's population. In Asia, one billion people depend on rice as a primary food source, while living in poverty and malnutrition. In Indonesia, rice is not only a food habit but also a strategic and political commodity that controls the lives of its people. Further, it has transformed into a strategic economic sector for economic and national food security. On average, rice consumption in Indonesia is 139 kg per capita per year, with a production of 41 million tons of rice (68 million tons of grain). In Thailand, rice production is 20 million tons, while rice consumption is 70 kg. In Vietnam, rice production is 17 million tons, while consumption needs are only 90 kg per capita annually. In contrast to Indonesia, these two countries have a surplus and can export rice to Indonesia. Indonesia often experiences food crises, such as the 1966 crisis when the price of rice reached 600 times. The 2003 food crisis led to the liberalization of the logistics agency's role, while the 2007–2008 crisis led to soaring rice prices. Up to this point, the fulfillment of national food needs has continued to rise owing to the increase in food consumption and population growth. In 2019, rice consumption of all types, including local, superior quality, and imported rice, averaged 1.374 kg per capita per week. When the COVID-19 pandemic struck, average consumption rose to 1,451 kg per capita per week. High dependence on rice food has caused the Indonesian state to import rice from Thailand and Vietnam. If this trend continues, there are fears that the country may struggle to provide sufficient food for its population. Therefore, this study aims to encourage food sustainability in Indonesia. The research methodology employs a qualitative paradigm, with data collection conducted through in-depth interviews and focused discussions with key stakeholders. The analysis was carried out in a descriptive-analytical manner. This study is novel as it establishes a local food system by introducing non-rice food banks

H. Warsilah (✉) · A. Wahyono
Center for Society and Culture, National Research and Innovation Agency of Indonesia, Jakarta, Indonesia
e-mail: henn001@brin.go.id

A. Wahyono
e-mail: ary318@gmail.com

M. Yoshizumi et al. (eds.), *Navigating Local Sustainability in Food, Community, and Innovation*, https://doi.org/10.1007/978-981-96-8963-7_3

in the region to reduce dependence on rice food and encourage the consumption of local food.

Keywords Local food systems · Non-Rice food banks · Food habits · Food sustainability

1 Chapter Summary

This study narrates the efforts aimed at building a local food system through the establishment of non-rice food banks at the village level, both within and outside Java. This is because Indonesia is dependent on rice as a staple food, evident in its high consumption of rice food. However, in certain regions, staple foods include non-rice foods, such as corn, cassava, sago, and tubers, which characterize local food diversity. This food diversity is supported by the population's food habits, which do not solely depend on rice food. Additionally, the local staple food contains sufficient nutrients; therefore, cases of malnutrition are rare.

The local food habits underwent change with the implementation of food policies and political changes that began in 1970 under the New Order regime, marked by the launch of a rice self-sufficiency program. Although the New Order's food policy successfully encouraged rice production in all regions of Indonesia, it led to a shift in local food habits from non-rice to rice-based foods. As a result, in just 20–30 years, Indonesia experienced a food crisis and had to import rice from other countries.

In several areas studied, people continue to consume non-rice local foods, such as Trenggalek in Java and Minahasa outside Java. These two regions consume non-rice staple foods, such as corn, cassava, sago, and tubers (*gembili* and *porang*). However, they are not malnourished. This practice facilitates the maintenance of local food sustainability. Currently, several countries require *gembili* and *porang* rice owing to their excellent nutritional value.

To establish a local food system and maintain food sustainability, we propose the immediate implementation of the non-rice food bank concept by local governments within and outside Java, and even in other developing countries. Research indicates that efforts to adopt the food bank concept can break the chain of farmer poverty and landlessness. Additionally, it ensures the continuous availability of food for families and reduces dependence on imported rice food, which often experiences an upward price trend.

The institutional structure of non-rice food banks is simplified as much as possible, thereby reducing the burden on the structures belonging to each organization. It helps farmers sell local food products, make local food savings, and obtain loans easily. The implementation of this concept encourages collaboration among village communities.

2 Background of Indonesia's Food Problems

Currently, the world is facing a food crisis, as reported by the UN Food Agency, who noted that global food prices rose by more than 14% in 2022 compared to the previous year. In December 2022, global food prices marked the ninth straight monthly decline. The Food and Agriculture Organization (FAO) reported that the international price of the most traded food commodity globally reached an average of 132.4 points. September 2022 compared to a revised 135.00 points for November. The previous November Figure was 135.7 points. In 2022, the benchmark index averaged 143.7 points, up 18 points or 14.3% from 2021, which was the highest increase in food prices since records began in 1990 (see Economy ecozone.com., 2023). The need for food has increased dramatically, as reported by the FAO. Rice food has been determined as a key food worldwide and is consumed by approximately 3 billion people daily. Rice is a staple food for 60% of the world's population. In Asia, one billion people depend on rice as a staple food, while living in poverty and malnutrition.

World food prices are rising due to several things, such as changes in global food consumption and production patterns, the conversion of food land to other purposes, the use of food commodities for energy, and the influx of imported food commodities.

Within a smaller scope, this change in consumption and commodities is related to family food security. In the past, most families' food needs could be supplied directly from the farm. However, today, staple foods must be purchased from traditional and modern markets, resulting in families losing control over meeting their food needs and ensuring nutritional quality and value.

Indonesia has the third largest population in the world, with a total population of 277.43 million people, approximately 42% of whom live in rural areas (BPS, 2023). Thus, fulfilling food security is imperative as it is a basic human right. To address this, the Ministry of Agriculture has realized the defense budget for 2022 through the State Revenue and Expenditure Budget, allocating IDR 92.2 trillion for national food security. This Figure decreased by almost IDR 7 trillion, down 6.86% compared to the previous year's allocation of IDR 99.02 trillion (Databoks, 2022). Subsidies in the food sector, including rice subsidies for people living in poverty, increased from IDR 6.4 trillion in 2005 to IDR 13.0 trillion in 2009. The demand for rice food continues to increase in line with population growth, as well as the need for more agricultural land, as shown in Table 1.

3 Crises and Food Policy in Indonesia

Indonesia often experiences food crises, such as the 1966 crisis when the price of rice surged 600 times, the 1998 crisis when Indonesia experienced a financial crisis that led to a food crisis, the 2003 crisis which prompted the liberalization of the logistics agency's role, and the 2007–2008 crisis that witnessed soaring rice prices.

Table 1 Demand forecast for rice and Indonesian agricultural land by 2030

Year	Total Population (Million)	Rice requirement (Million Tons)	Land Requirements (Million Hectares)
2007	220	30,58	11,6
2011	237, 641	33,32	13,566
2022	277,43	54,5	13,17
230	425	59	23,4
	Lack of Rice	26,66	9,9

Source: Coleman, (1981), directorate general of land and water management

In addition to the food crisis, the fulfillment of national food needs continues to rise owing to the increase in food consumption and population growth. Agriculture has been developed to address the fulfillment of national food. However, the results have yet to impact people's welfare and have failed to change the conditions of poverty and food insecurity in rural areas.

The consumption patterns of the Indonesian population are increasingly dependent on rice as a food source, although the tradition of consuming non-rice foods persists. The high per capita consumption of rice is linked to the national rice stabilization policy, wherein the state actively regulates the availability of rice food between seasons. Through its national logistics agency, the state distributes rice meals between seasons and during famines through market operations. The state immediately sets a floor and fixed price for rice, aiming to protect producers from dramatic drops in rice prices.

Additionally, agricultural revitalization in non-rice producing areas has proven successful. It has created a market for non-rice commodities originating from imported seeds, such as hybrid corn, offering an income for farmers who have thus far depended on perennials such as coconut/copra. For example, in northern Sulawesi, superior (hybrid) maize varieties are valued as commercial crops that provide income to household farmers.

The success of agricultural revitalization has resulted in a decrease in local corn food production, although it functions as a non-rice local food. Local corn cultivation has several advantages.

a. It has the taste and smell of sweet corn.
b. Lasts a long time when stored.
c. It tastes good, particularly when used to make *Milu Rice* or cake snacks.
d. The corn grains are large.
e. It requires minimal maintenance during planting.
f. Post-harvest is easy, involving sun-drying (Warsilah, 2013).

Meanwhile, the high-yield varieties of corn introduced by the government are crops whose production results are intended for sale as animal feed (for pigs and dogs) and are not considered a food reserve. Regarding food security, this superior variety of corn is not a local food source.

The agricultural revitalization program (maize) is a government policy that encourages the productivity of agricultural products by increasing farmers' income. The development of high-yield maize varieties is killing local maize varieties and creating challenges to obtaining local food sources. The success of the corn revitalization program has greatly affected the availability of local food by reducing the production of local corn varieties. Rice remains an expensive food for people in rural areas living in poverty, making it unaffordable. Additionally, the price of rice fluctuates and rice can disappear from the market at any time. The authors analysis predicts that this will result in a food crisis, as the local food system will no longer support people's food needs.

The high dependency on rice food consumption in rural areas leads to a large allocation of household income to purchase rice food. Additionally, Indonesia is nationally highly dependent on carbohydrate food sources such as rice and wheat/wheat flour. This is evident in efforts to import rice and carbohydrate foods exceeding the government-set quota by more than 62%. The country's dependence on rice food reflects people's food consumption patterns, which tend to be rice, despite the abundance of non-rice food sources, such as cassava, sweet potato, corn, soybeans, bananas, sago, and other tubers.

See table Fig. 1 below, in the research area, in Minahasa, North Sulawesi, local corn production, or *milu*, is very high. Between 2012, production reached 21,500 tons and fell to 7,300 tons in 2015. Local corn is a local non-rice food ingredient which is the food habit of the Minahasa population.

Table 2 shows the level of cassava production in the Trenngalek area in the 2017–2018 period, producing more than 100 tons. Residents process this cassava into *tiwul rice*, as a substitute for rice. Trenggalek residents consume processed cassava *tiwul* as a daily substitute for rice (Fig. 2).

The shift towards a dependence on rice food was driven by the politics of rice self-sufficiency in Indonesia in the 1970s. This gradually changed the food habits of Indonesian people in rural areas from non-rice food to rice food. In response to these conditions, the government is encouraging food diversification to reduce the burden of consuming carbohydrates to other cheaper and more affordable food commodities.

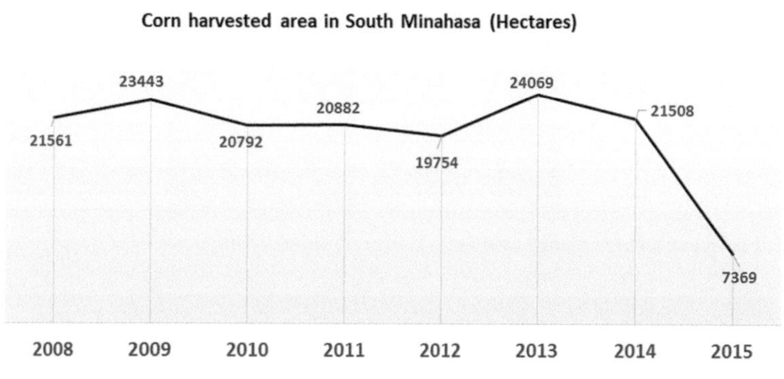

Fig. 1 Maize production in South Minahasa, 2008–2015 *Source* BPS North Sulawesi (2020)

Table 2 Local food diversity: non-rice food

No	Area	Non-Rice Food	Local language	Image of Non-Rice Food	Image of processed Non-
1	Trenggalek - East Jawa	1.Rice Casava 2.Gembili	Tiwul 	 	 
2	South Minahasa North Sulawesi	1.Corn Rice 2. Banana	Milu Banana frite		
3 .	Bahoi and Bulutui- North Sulawesi	1.Corn Rice 2.Rice Casava 3.Banana.	Banana frite		
4	Kalimantan and Ternate- North Maluku	1.Rice Casava 2.Sago			
5	Current food choices in urban areas	Porang Tubers			

Source LIPI/BRIN research results (2013)

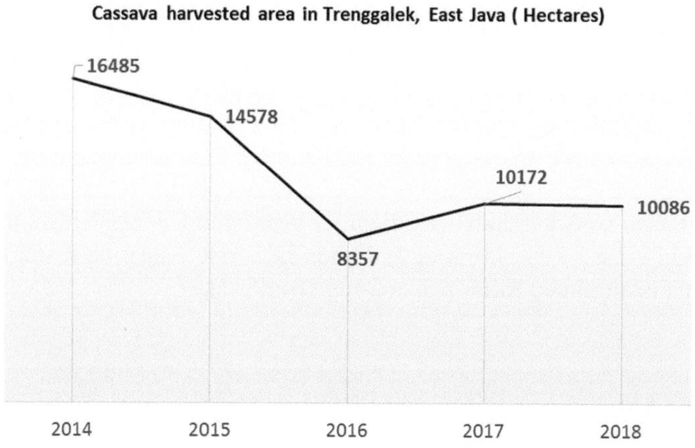

Fig. 2 Cassava production in Trenggalek, 2014–2018 *Source* BPS East Java (2020)

The Indonesian government's policy has directed efforts toward strengthening food security based on the self-reliance and diversification of domestic production, and the creation of a favorable farming business climate, and the maintenance of agricultural development sustainability. This policy aims to establish food security nationwide. Building food security and sustainability is crucial and strategic for nourishing the community. It encourages food sustainability by developing a food production system based on local resources, food institutions, and culture.

The results of the Indonesian Institute of Sciences' (LIPI) research conducted in 2009–2013 in Trenggalek Regency, East Java, South Minahasa, and North Sulawesi, stated that not all individuals immediately switched to food rice as their staple diet. The practice of consuming non-rice foods such as corn rice, cassava rice, or sago flour is a part of the local community's food habits. The persistence of non-rice food consumption practices is also influenced by the unequal economic capacity of individuals, distance between the countryside and the city center, and local food habits. According to Novianti et al. (2013), not all rural communities can afford to purchase rice; some buy rice only as an ingredient for mixed food, like "corn and rice." The rice assistance for people living in poverty was distributed unequally. Therefore, a consensus among villagers led to an equal distribution of rice assistance. Thus, each household receives only a small share. This is because rice can only be used to produce mixed rice (rice mixed with corn). However, nationally, Indonesia remains highly dependent on rice as a staple food.

To overcome the problems of a food crisis, the authors propose the establishment of non-rice food banks through research conducted in various regions of Indonesia, to reduce dependence on rice food and restore local food wisdom. According to Warsilah (2011), if the dependence on rice food continues to be allowed, instead of proclaiming "local food sovereignty," certain areas with cultural wealth in the form of non-rice staple local food habits will, in the short term, increasingly depend on rice as a staple food. It is certain that the area will experience a deficit in rice food availability, thus threatening food insecurity. To restore local food sovereignty, various policy support and programs that encourage the consumption of non-rice staple foods should be implemented in regions with these food habits. This can be achieved by establishing non-rice food banks. This ensures that the government can continue to provide for the community and maintain food sustainability.

The important research questions to be answered in this study are: "How are regional governments and local communities overcoming dependence on rice food and the food crisis? Can efforts to build non-rice food banks be realized in rural areas? What institutional forms can non-food banks take?

4 Methodology

This study uses a combination of quantitative and qualitative research methods. For the quantitative aspect, questionnaires were administered to 100 respondents in each region, targeting households living in poverty, comprising farmers, planters, laborers, and other food-insecure households. Questionnaires implies that several questionnaires were distributed, consisting of personal data of respondents, information on agricultural and plantation products at the household level, details on food habits, information on food security, local food and its processing, and barriers to family food procurement.

To gather qualitative data, in-depth interviews and focus discussions were conducted with several important stakeholders who understand local food, food habits, and local food security. Data collection was also carried out through observation, recording, and conducting focused discussions and in-depth interviews with the Department of Agriculture, Fisheries, Forestry, Food Security, Statistics, Regional Planning, and Regional Government Offices. Questionnaire data underwent statistical processing, whereas qualitative data were nalysed using analytical description. Research locations were Java (Trenggalek, East Java) and outside Java (North Sulawesi).

Although data collection using a combination of qualitative and quantitative methods was carried out in 2013, with subsequent updates in 2020 in Trenggalek and South Minahasa to observe people's food habits, there was minimal change from the initial results. The community continues to consume non-rice local foods such as corn rice (*milu*) and cassava rice (*tiwul*). Rice food is typically consumed during celebratory events or on weekends, often mixed with *milu* or *tiwul*. Therefore, we consider the 2013 data as relevant and valid.

5 Framework for Food Security, Food Habits, Food Diversification, and Non-Rice Banks

5.1 Food Security

Food is a basic need for every human being. Based on Law Number 7 of 1996, food is defined as something that comes from nature and can be consumed. The fulfillment of food needs is important in the context of national development, serving as a benchmark for community welfare (Hanafie, 2010). Food crisis conditions in a country have a food crisis in a country economic (inflation), social, and political stability. In Indonesia, the term food security has been known since the 1970s, focusing only on the supply of self-produced food at affordable prices. By the 1980s, the concept expanded to include access to food per household. In 1990, food security was defined as the affordable production of food, considering environmental friendliness (Ariani, 2002).

According to the FAO, food security extends beyond fulfilling people's needs for nutritious and high-quality food. It involves ensuring the availability of food to meet consumption needs. Food must be available under any condition (stability) and access to food, namely, the community's ability to produce or purchase the food they need, is essential (Hanafie, 2010). Food security comprises several subsystems, including food availability, access to food, and nutritional status (Rustanti, 2016). All these subsystems must be fulfilled; if any one of these aspects is not met, food security cannot be ensured. As quoted by Suhardjo (1996) in Ariani (2002), the 1984 FAO conference established the principles for food security, guaranteeing adequate availability of food for humankind and ensuring that every individual could obtain food. The definition of food security was refined during the International Congress of Nutrition (ICN) held in Rome in 1992. Household food security is a household's ability to consistently provide adequate food for its members, enabling them to lead healthy lives and carry out their daily activities. However, in the 1995 Committee on Work Food Security meeting, this definition was expanded by incorporating the condition that the requirements must be accepted by the local culture (Soetrisno, 1997). This definition was reaffirmed in the Rome Declaration on World Food Security and the 1996 World Food Summit Follow-up Plan to become "food security exists when all people, at all times, have physical and economic access to sufficient, safe and nutritious food to meet needs according to their tastes for an active and healthy life."

Indonesia, one of the countries that has expressed its commitment to implementing the Rome Declaration, accepts the concept of food security, which was legitimized in the formulation of Food Law No. 7 in 1996. Food security in Indonesia includes aspects of safety, quality, and diversity as conditions that must be met to meet the food needs of the population in an adequate, equitable, and affordable manner.

Indonesia's food security system includes: (1) the quantity and type of food available that is sufficient for the community; (2) equitable distribution of food among all people; (3) consumption of food with balanced nutritional value; and (4) nstalment of the community's nutritional status (Achmad, 2014). As Reutlinger (1987) revealed, food security can be interpreted in several ways. Braun (1992) also revealed that the term food security can lead to debate and confusion owing to its broad scope. Nevertheless, it remains a crucial concept for many people globally. Food security can be realized by establishing a local food system oriented toward resource utilization and meeting local needs, starting from the family and extending to the community level. Yards are an important part of this system. The systems for growing food have long been used in central Java and other regions. Thus, food security, based on local wisdom, can be used as an alternative solution to increase Indonesian food security, starting with family food security. The nstalment of a country's food needs has an impact on the quality of human resources residing in that country and will have implications for the welfare of society.

5.2 *Food Habits*

Dietary patterns are the ways in which a person or group chooses food and consumes it as a reaction to physiological, psychological, cultural, and social influences (Harper et al., 1986, p. 251). Food patterns can foster food habits in society because they are human behaviors that fulfill the need for food, and includes attitudes, beliefs, and food choices (Khumaidi, 1989:27). Food habits can be defined as the entrenched way of eating adopted by an individual or society; food habits that exist in a community in an area may differ from another. Suhardjo (1989:20) emphasized that food habits are both a cultural and social symptom, offering an overview of the behavior and values upheld by an individual or group. Cultural differences in each society determine differences in staple foods.

Based on this, Koentjaraningrat (1984:34) developed a model to examine the social and cultural factors influencing dietary and family food consumption patterns. Accordingly, the diets of individuals, families, and communities are influenced by the following factors. First, cultural factors, including what people think, feel, and perceive about food. Thoughts, feelings, and views (perceptions) are expressed through actions (what people do and practice) such as eating and making food choices. If this mechanism is repeated, consumption ehaviour becomes an eating habit that can be measured. Second, food "consumption pattern" factors can be observed and measured. Third, social environmental factors (population aspects such as composition, strata, and characteristics), ecological environment (soil and climate conditions), and biological/economic environment (farming and market systems). Fourth, cash availability is an important factor.

People's attitudes toward eating can be either positive or negative and are influenced by values that are directly experienced; individuals' preferences stem from external factors (Khumaidi, 1989). The external factors influencing eating patterns are as follows: (1) the natural environment; (2) people's food patterns; (3) the social environment, which provides a clear picture of differences in eating patterns. Dietary practices vary among nations and ethnic groups, based on the culture that has been passed down from generation to generation; (4) cultural and religious environments, where diets usually embody the values of spiritual life and social obligations; and (5) the economic environment and food distribution, largely determined by community groups based on their economic levels. Thus, eating habits are sociocultural values that are passed down from generation to generation and shape the ehaviour, attitudes, and beliefs of a person or group who refer to local cultural references. The food habits of each community group in an area, as well as their food needs and consumption patterns, differ from those of other community groups. However, coercive external factors, such as policy factors (rice self-sufficiency), economic factors (buying power and food distribution), environmental factors (environmental changes), and factors related to harvest and post-harvest technology, cause insufficient food availability, thereby leading to shifts and changes in people's staple food habits.

5.3 *Local Food Diversification*

Food diversification, which is now a priority for the government to achieve food security, requires commitment and new breakthroughs to reduce dependence on rice. Based on Warsilah's (2013) research presented at National Geographic, food supply and production are affected by climate and global environmental changes. There is an imbalance between production and food stocks between regions, and variations in mastery and the ability to apply food production and processing technology. This is coupled with the degradation of land and water quality and environmental damage. The complexity of the global food security problem also concerns food demand and consumption. In several regions worldwide, the population continues to grow, particularly in Africa and Asia. Additionally, there is competition for food utilization between food feed fuels. Warsilah also explained that domestic food problems are no less complicated. As stated by Nicholas Stern in the Review on the Economics of Climate Change (2007), developing countries are already experiencing food insecurity in the context of climate change. Developing countries often possess limited capacity to address the effects of climate change. One of the most serious impacts of climate change is experienced by the marine sector, affecting the food conditions of inhabitants in coastal areas. According to Warsilah, the urgency of environmental changes and their impact on social aspects becomes evident when they interact to exacerbate food, social, economic, and political crises. This also increases poverty, inequality, and food vulnerability.

Food diversity or diversification, observed through people's food habits, aims to broaden people's choices in consumption based on individual taste preferences, avoiding monoton in obtaining food and nutrition to live a healthy and active life.

Implementing the concept and operation of food diversification and related policy tools poses a dilemma for policymakers at the national level. However, to date, food diversification has not been effective (Harper et al., 1986). The guiding principles to encourage food diversification are as follows: (1) from the consumption side, food diversification is not intended to replace rice but to change people's food consumption patterns; (2) on the production side, food diversification is a process of adjustment to reduce risks, increasing the income of farmers, and anticipating long-term demand; and (3) the smooth operation of food diversification is achieved when combined with the development of the agro-industry in rural areas through vertical integration with food crop farming. Another crucial principle is that food diversification is neither a goal nor a policy instrument to achieve the stabilization of rice. Instead, with the development of the agro-industry in rural areas, it is expected that achieving macro-scale goals (national interest) can go hand in hand with achieving micro-scale goals (farmer interests) (Harper, 1986, and Krisnamurthi, 2003).

According to Warsilah (2013), food diversification is the process of choosing a variety of foods rather than depending on one type. This encompasses the production, processing, and distribution aspects of food consumption at the household level. Food diversification was key in successfully developing agriculture during 2010–2014 and was an important aspect of the performance contract between the Minister of Agriculture and the President. The Minister of Agriculture's work contract with the President of the Republic of Indonesia included four key achievements: (1) increasing self-sufficiency and sustainable self-sufficiency; (2) increasing food diversification; (3) increasing value-added, competitiveness, and exports; and (4) increasing the welfare of farmers and fishermen. However, Warsilah (2011) found that the Indonesian government's food diversification program is often constrained by myths circulating in society. One such myth involves the consumption of tiwul (rice cassava), suggesting that it can affect public education. This assumption should be dispelled, as there is no evidence that individuals consuming *rice tiwul*, a local food, experience a decline in education levels because of its negative impact on their health. Warsilah, having conducted research in two locations, concluded that areas where rice *tiwul* is consumed exhibit better public health levels and higher life expectancies, with most people surpassing 80 years of age. The study's results indicate that many consider *rice tiwul* consumption to provide increased physical strength. Further, they believe that *rice tiwul* consumption leads to a form of addiction, asserting that *tiwul* has become an opium for the body—a major need that cannot be replaced even by rice (see c8/pr/jatimprov).

Warsilah further stated that efforts to restore local food sovereignty require broad and continuous public participation. This is because dietary patterns serve as a means for a person or group to choose and consume food in reaction to physiological, psychological, and sociocultural influences. Food patterns can also be interpreted as manifestations of human nstalmen or a group's nstalment of its food needs, encompassing attitudes, beliefs, and food choices. As food habits or eating patterns are part of society's culture, the involvement of all parties is required to ensure local food sovereignty.

The Indonesian Food Security Agency, through the Ministry of Agriculture, implemented a food security development strategy that encourages increased consumption of diverse, nutritionally balanced, and safe local foods to reduce rice consumption, with a target of 1.5% per capita annually. Indonesia's dependence on carbohydrate food sources, such as rice and wheat flour, as well as protein food sources, remains high. For example, in 2007, grain imports reached 62.2% exceeding the government-set target of 50%. Meanwhile, imports of non-rice food groups, such as new tubers, accounted for 3.1% of the recommended 6%, while vegetables and fruits comprised 5% of the recommended 6%.

The establishment of a local food system is expected to safeguard a community's food reserve system. Developing a local food reserve system guarantees a steady and sufficient supply of food for all villagers, particularly during rice scarcity and rising prices. The LIPI, through Wahyono (2011), conveyed the importance of establishing non-rice food banks to build a local food system, with the aim of strengthening the basis for realizing food sovereignty in rural areas. With local food sovereignty, a community can determine the type and amount of food to be produced, consumed, and traded.

Table 2 below describes a variety of local food diversity, which is non-rice food. There is a variety of local food, starting from cassava which is an ingredient in tiwul, porang tubers, corn or milu and bananas. This table also shows the various forms of processing from boiling, fermenting, and frying, thus illustrating the richness of local non-rice foods.

The types of staple foods, other than rice, in the locations studied, both in Java and outside Java, are highly diverse. In South Minahasa, Bulutui, and Bahoi-North Sulawesi, the population consumes local food, such as corn rice, known as *milu rice* in the local language, and bananas, which are prepared by frying. In Trenggalek, East Java, staple foods include cassava rice, also called *tiwul rice*, as well as *gembili* and *porang* tubers, which have been processed into *porang rice* or *gembili rice.* In Kalimantan, people in rural areas consume sago, cassava and palm, which are sliced by roasting and then processed into sago porridge. Currently, there are new types of non-rice foods gaining popularity and even exported to Japan, namely *gembili* and *porang* tubers. *Porang rice* has significantly fewer calories than regular rice. It contains many fibers; however, they do not derive from starch. Shirataki contains approximately 97% water and 3% glucomannan fibers. This makes shirataki lower in calories; therefore, it is believed to help maintain a healthy diet. In one serving of *porang rice*, or the equivalent of 40 g, there are as many as 28 kcal of calories after cooking. The carbohydrate content is only approximately six grams and the protein content is approximately one gram. *Porang rice* can be used as a substitute for rice because of its low-calorie content. In comparison, one plate of cooked white rice contains at least 204 cal (https://seruni.id/menmenal-beras-porang-serta-benefits/).

5.4 Establishing Non-Rice Food Banks

The development of local institutions in rural areas benefits the food security of communities through mutual support and reinforcement. The establishment of non-rice food bank institutions aims to build a local food system to strengthen the basis for realizing community food sovereignty at a higher level. This allows communities to determine the type and amount of food they produce, consume, and trade. Building a local food system is a reasonable choice for preserving the food reserve system developed by society. It can guarantee the sufficient fulfillment of food for all villagers at all times, particularly during rice scarcity or when rice becomes unaffordable. Through food storage and cultivation of food reserve plants, food bank institutions can be formed to overcome food security challenges in rural communities.

An institution is often understood as a social institution or organization (social organization). These two terms can be distinguished. First, institutions tend to be traditional, whereas organizations tend to be modern. Second, institutions tend to be formed by rural communities, while organizations are generally formed by urban communities (Tjondronegoro, 1999). These definitions are not categorical. Institutions and organizations exist on a continuum, and the ultimate goal of an organization is to become an institutionalized organization or an institution with organizational aspects (Norman Uphoff, 1986). Initially, the terms "institution" and "organization" tended to be indistinguishable and were sometimes used interchangeably.

Esman and Uphoff (in Uphoff, 1986) distinguished institutional forms at the local level as follows:

a. Local administrative institutions comprise local agents and government officials operating at the local level. These government employees are accountable to the government agency that assigned them.
b. Local government is an executing institution elected or appointed by the local community and has the authority to handle development and regulatory tasks. This institution is responsible for the local communities.
c. Organizations, local associations, and institutions whose members aim to handle a variety of interests, typically related to members' needs and specific tasks, such as water use.
d. Cooperative institutions, such as cooperatives, are local organizations that work by pooling the resources of their members for mutual benefit.
e. Service organizations are local organizations formed to assist non-members, although members may also benefit from such organizations, for example, Red Cross.
f. Private businesses encompass independent businesses or branches of institutions outside the local area. An institution is a business unit involved in production, service, and trading activities.
g. Trade.

6 Discussion of Research Findings

6.1 Description of Initiative Activities: Efforts to Strengthen Sustainable Local Food by Establishing Non-Rice Food Banks

The study, conducted by the Center for Research on Society and Culture, Indonesian Institute of Sciences (PMB-LIPI) team, began in 1998 during a financial and monetary crisis in the country that led to a food crisis. As a response to the food crisis conditions in Indonesia, action research was carried out regarding the establishment of a rice food bank in Indramayu Regency, West Java." Then, the study continued from 2009–2013 and focused on creating an institutional model for non-rice food banks in Trenggalek Regency, East Java, and South Minahasa Regency, North Sulawesi. This LIPI intervention analyzes the implementation of food security policies to achieve food self-sufficiency or food sovereignty at the community level.

The institutional non-rice food bank is a community empowerment program that restores community authority for protecting, conserving, and utilizing agricultural areas and other environments to ensure the nstalment of food needs and income. The non-rice food bank is a social institution that takes various forms such as cooperative institutions, service organizations, local organizations, and private businesses. However, food banks, as an institutional entity, are inseparable from the structural and cultural aspects of institutions. The development of non-rice food bank institutions involves building a local food system to strengthen the basis for realizing village food sovereignty through nutritional nstalment.

Corn is an important source of protein. The main nutritional content of corn is starch (72–73%), composed of amylose and amylopectin in a ratio of 25–30%: 70–75%, except for waxy corn, which has a ratio of 0–7%: 93–100%. The sugar content of corn (glucose, fructose, and sucrose) ranges from 1–3%. Corn protein (8–11%) consists of five fractions: albumin, globulin, prolamin, glutelin, and non-protein nitrogen. In addition to functioning as a micronutrient, corn also acts as a natural antioxidant that boosts immunity and inhibits cellular degenerative damage. Additionally, corn contains various essential minerals, such as K, Na, P, Ca, and Fe (Calcium (Ca), Phosphorus (P), Sodium (Na), Iron (Fe), phosphorus (P)). Genetic factors significantly influence its chemical composition and functional properties (Suarni and Widowati, 2016). Cassava contains various important nutrients. In 100 g of boiled cassava, 98% of calories derive from carbohydrates and the remainder from protein and fat. At the same serving, cassava contains fiber, vitamins, and minerals. The nutritional content of 100 g of cassava is as follows: water: 61.4 g; carbohydrates: 36.8 g; energy: 154 cal; protein: 1.0 g; fiber: 0.9 g; fat: 0.3 g; potassium: 394 mg; calcium: 77 mg; vitamin C: 31 mg; and phosphorus: 24 mg (https://hellosehat.com/nutrients-facts-nutrition-dan-benefits-cassava/).

The food bank in Minahasa stores corn in the form of ground dry corn, which is easy to market and store. Meanwhile, the cassava harvested in Trenggalek is stored in its original form, soaked for several days, dried, and then cut into small pieces. In this preparation, tiwul can be made as a substitute for rice. These two villages consume their local food daily, except on holidays when local food is mixed with rice. The local food and rice are cooked separately; after cooking, they are mixed together. Consumption of local food is usually accompanied by processed vegetables in coconut milk, fried sea fish, or salted fish. Based on the study's findings, Minahasa and Trenggalek residents have long life expectancies, and the older population remains sufficiently strong to perform gardening activities. Their children attend school up to the high school level, indicating that they receive good nutrition and rarely suffer from illnesses.

The results of the 2013 LIPI/BRIN research also highlighted that the culture of non-rice food habits was spread to urban areas with the slogan "One Day No Rice," which attracted younger people. The success in encouraging a reduction in rice consumption in urban areas is associated with information emphasizing that excess consumption of rice can cause diabetes, owing to its sugar content. Additionally, a local policy was implemented, stating that individuals hosting celebratory parties and inviting officials must serve food processed from non-rice food. This policy encourages public interest in processed non-rice-based foods and increases local food consumption.

Cassava is a local food ingredient used to produce *tiwul rice*. The process of making *tiwul* is simple, albeit lengthy, as the cassava must be dried first. Before processing, the cassava is cleaned, cut into small pieces, and sun-dried. The drying time depends on the weather and is usually approximately five to seven days. Once dried, the cassava is ground or pounded with a pestle to produce flour, which is processed by adding a little hot water, followed by kneading and steaming until cooked. After *tiwul* is cooked, it is usually referred to as cassava rice or *tiwul rice* (see Kompas.com, 2021). Corn rice, made from corn, has a distinctive taste and aroma. The production method, considered relatively easy, involves drying corn kernels in the sun or over a cooking stove, then coarsely grinding them into small grains. Before processing into corn rice, the mashed corn grains are soaked in water, drained, and cooked by steaming. In addition to its distinct taste and aroma, corn rice has many other benefits. Among them, 100 g of corn provides as many as 342 cal and contains B vitamins. It is good for consumption by pregnant women, as corn rice contains a high amount of folic acid, helping to prevent neurological defects and fetal weight deficiency.

The Fig. 3 below is a form of rice cassava or *rice tiwul*, the brown one is *tiwul* filled with sugar so that it functions as a cake snack. Meanwhile, in the picture on the side, *tiwul rice* is used as a substitute for rice, and the way to eat it is using vegetable soup.

Figure 4 above is corn rice or *milu rice*. Meanwhile, in the side picture, *milu rice* has been mixed with rice and the way to eat it is using vegetable sauce.

Fig. 3 Rice cassava "*tiwul*"

Fig. 4 Rice Corn "*milu*" and rice mixed with corn *Source* Warsilah et al. (2013)

According to LIPI findings, it is necessary to establish non-rice food banks to encourage non-rice food security. Food banks can also serve as a revitalization platform for combined farmer groups (Gapoktan). Gapoktan is more appropriate as a development agent, because it must emerge from exclusivity as a medium for institutional transformation. Gapoktan also spearheads agricultural services. Non-rice food banks are limited to strengthening horizontal ties, namely membership bonds with shared activities, to establish cooperation and togetherness. Subsequently, they can be upgraded to an organization that is legal, modern, and independent. The institutional development of non-rice food banks will not be positioned as an "institution as a project tool," but as a form of community empowerment and to achieve local food self-sufficiency. However, it is important to note that efforts to build food security at the community level through the establishment of non-rice food banks are futile without first addressing farmer poverty.

This action research aims to explore methods to reduce the number of agricultural laborers and landless farmers, as well as to build food security. Efforts to build a local food system are a rational choice because they can maintain a food reserve system at the community level. By establishing food bank institutions and cultivating non-rice food reserve crops to ensure food security in rural communities.

This action research aims to explore methods to reduce the number of agricultural laborers and landless farmers, as well as to build food security. Efforts to build a local food system are a rational choice because they can maintain a food reserve system at the community level. By establishing food bank institutions and cultivating non-rice food reserve crops to ensure food security in rural communities.

6.2 Activity Output and Outcome: Partiality to Economically Disadvantaged Rural Farmer Groups and Encouraging Facilitators

The initiative to build non-rice food banks at the community level for farmers living in poverty in the studied areas began by providing capital to landless farm laborers, enabling them to participate in field or garden auctions. Successfully rented paddy fields and gardens were distributed to farm laborers. Every farm worker was expected to cooperate in cultivating rice fields, from processing to post-harvest, and work on the rice fields together in turns. Food banks were established to address food insecurity, particularly during famines. This involved setting aside a portion of the collective harvest yields. After the farm laborers obtained their leased land from the government, called "Pangonan" in the local language, communal cultivation of the fields was carried out, and a portion of the collective harvest was set aside for the food bank. The communal approach to cultivating paddy fields and gardens, as well as establishing a food bank, is considered more effective than an approach that views farmers as individuals. In the individual approach, the social context of farmers is often ignored.

The management of the food bank was established to include five board members selected from the following groups: community leaders, village officials, and representatives of researchers/village government; three individuals who were designated the roles of chairman/manager, secretary, and operational officers responsible for procurement and distribution; and two field coordinators to oversee land credit, food, and fertilizer availability. While managers do not receive a salary, they are entitled to incentives. The incentive system is as follows.

The board receives a 2.5% incentive at the end of each harvest season, obtained from food bank revenue (50% of crop yield) and allocated for land lease and fertilizer credit nstalment. The incentive is calculated after deducting the amount reserved for land and fertilizer credit expansion of 1 ton of GKP (equivalent to 666.67 kg of rice/tubers/corn) per hectare. Food bank managers, excluding the board of directors, collectively receive a 2.5% incentive per month based on the number of commodities nstalment as food loans. Additionally, managers are entitled to 2.5% of their credit nstalment receipts. Villagers eligible for food bank membership include heads of families of farm laborers and small farmers with arable land of fewer than 0.1 hectares.

Each food bank member is mandated to pay a monthly fee of Rp. 1000 for additional revolving fund capital or food bank operational costs. The management structure of a non-rice food bank is illustrated in the Fig. 5.

Commodities prioritized in food procurement are corn, cassava, potatoes, and sweet potatoes, adapted to local food habits. The excess from the harvest from shared land or "Garapan Bersama's land" will be sold to the market, and the money will be used for members' savings and loan cooperatives.

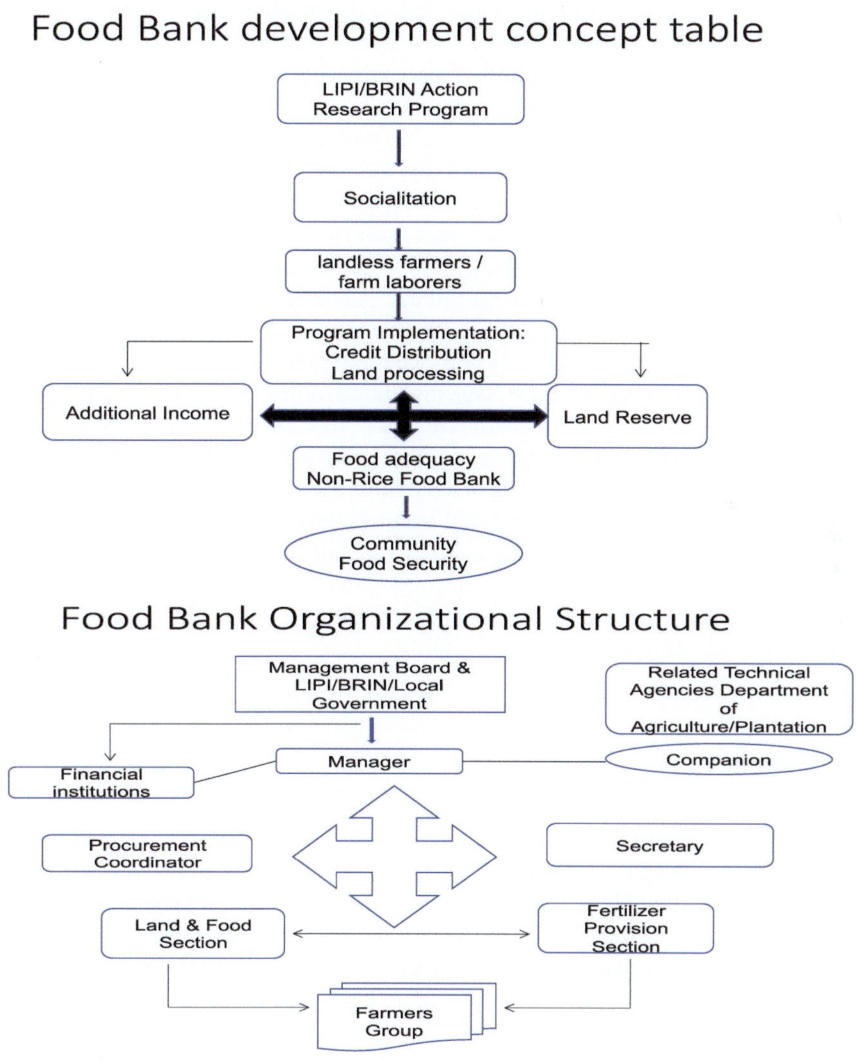

Fig. 5 Food bank development concept *Source* LIPI Research Results (2013)

The main duties and functions of non-rice food banks, based on an agreement between the community and management, are as follows: (1) establishing cooperation with various stakeholders to ensure the availability of non-rice food; (2) supervising food reserves in individual gardens, carried out by farmers, farmer groups, extension officers, sub-district agricultural officers, and village heads; (3) supervising food reserves in collective gardens, carried out by the village head, farmer groups, and District Agriculture Mantri; (4) supervising at the district level, carried out by the Agriculture Office, the Local Food Sovereignty Council, and the Local Food Security Agency; (5) introducing agricultural technology (seeding, fertilization, cropping patterns, and harvest-handling technology); (6) post-harvest technologies by: (drying, storing, cutting, splitting, preserving, and monitoring); (7) facilitating marketing production services both in the form of semi-finished and finished goods (cakes/various types of food). The institutions involved are, for example, the Department of Industry and Trade, food bank, Tourism Office, food industry, and Market Service; (8) managing distribution and marketing, which can be conducted by farmer groups and the Department of Industry and Trade; and (9) engaging in public advocacy. These activities include strengthening local food habits and advocating for the community and local governments through public campaigns integrated into the school curriculum.

Figure 6 below is the structure of a non-rice food bank developed by local communities. Based on the Non-Rice Food Bank institution formed by this community, Food Credit is distributed to members who have urgent food needs. The loans distributed are in the form of rice, corn, cassava, and sweet potatoes according to eating habits. Each member can borrow up to 30 kg of local food. Members can borrow a maximum of two consecutive months until the total loan reaches 60 kg of local food. The loan must be repaid within one planting season (four months) or during the harvest season. The lending mechanism is carried out through groups, field coordinators, and food bank managers. To enforce membership discipline, every loan repayment contains sanctions in the form of shared responsibility.

7 Conclusion

Currently, in response to the intervention of the LIPI, the Minahasa Selatan-North Sulawesi and Trenggalek regions in East Java have implemented the non-rice food bank concept to strengthen food security and preserve local food habits. Minahasa Selatan, with the local food corn (Milu in the local language), and in Trenggalek, with the local food cassava rice (known as Tiwul), have successfully promoted sustainable food. This indicates that efforts to build population food security by establishing non-rice food banks have proven to provide nutrition for the community and ensure sustainable maintenance of food. Thus, if each region adopts and establishes non-rice food banks, it will reduce Indonesia's dependence on rice imports from other countries. Other implications include reducing poverty and landless farmers, and the maintenance of local food sovereignty.

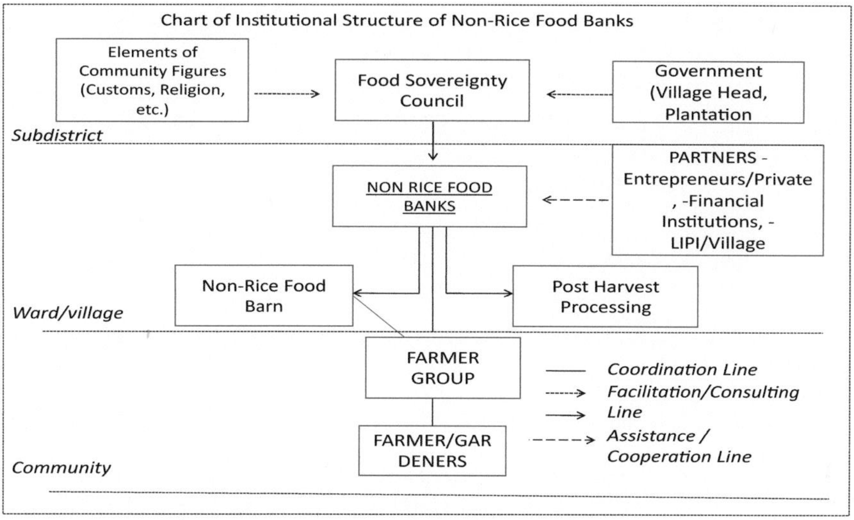

Fig. 6 Institutional structure of the food bank on rice *Source* LIPI Research Results (2013)

8 Recommendations

(1) Local governments throughout Indonesia should adopt the concept of non-rice food banks to reduce rice food insecurity and food crises. This approach may be applied to other developing countries with food insecurity problems.
(2) Non-rice food banks can strengthen food sovereignty and preserve local food to maintain food sustainability.
(3) It is important to educate communities so that the nutritional content of local food is sufficient for their nutritional needs, eliminating concerns about malnutrition.
(4) Regional governments must take an active role in marketing non-rice food products from their regions, ensuring proper packaging for distribution to other regions within Indonesia and even overseas. This could help farmers living in poverty keep food banks operational. In short, if successful, non-rice food banks could break the cycle of poverty and malnutrition in society.

References

Anonymous. (2001a). *Food awareness development work program. center for food alertness* 2001–2004.
Anonymous. (2001b). Program kerja pengembangan kewaspadaan pangan. *Pusat Kewaspadaan Pangan* 2001–2004.
Ariani, M. (2002). Ketahanan pangan: Konsep pengukuran dan strategi. [Unpublished manuscript].
Badan Pusat Statistik. (2023). *Data BPS 2023*. BPS.

Braun, VJ., Bouis, H., Kumar, S., Pandya-Lorch, R. (1992). *Improving food security of the poor: concept, policy and programs*. IFPRI, Washington, DC.

Coleman, J. S. (1981). Social capital in the creation of human capital. *American Journal of Sociology, 94*, S95–S120.

Databoks. (2022). https://databoks.katadata.co.id/datapublish/2022/04/20/anggaran-ketahanan-pangan-2022-dipangkas-hampir-rp7-triliun.

Economy ecozone.com. (2023). https://economy.okezone.com/read/2023/01/06/320/2741628/harga-pangan-dunia-Print-rekor-tertinggi-sepanjang-2022

file:///Users/henny/Downloads/143-Article%20Text-277-1-10-20150921.pdf

Husken, F. (1998). *Village Communities in Social Change in Java in 1830–1980*. Publisher PT.

Harper, I., Deaton, B. J., & Driskel, J. A. (1986). *Food, Nutrition, and Agriculture*. University of Indonesia Press, Jakarta.

Handewi, P. S., & Rachman. (2002). Ketahanan Pangan: Konsep, Pengukuran Dan Strategi. FAE. *Juli, 20*(1). 12–24.

Hanafie, R. (2010). *Introduction to Agricultural Economics*. ANDI Publisher.

Hanani, Nuhfil. 2009. Food Security in East Java. (http://docs.google.com). Measurements and strategies. FAE. *20*(1), 2002: 12–24

Jerome et al. (1986). *Nutritional anthropology: contemporary approach to diet and culture.*

Khumaidi, M. (1989). *Community Nutrition Teaching Materials*. Center Inter Univ. Food And Nutrition.

Khrisnamuti Bayu. (2003). Food Diversification and Indonesian Food Politics. *Proceedings of the Seminar series on Strengthening the Indonesian Food Diversification Roadmap, IPB* https://repository.ipb.ac.id/jspui/bitstream/123456789/52761/1/bayu%20krisnamurthi%20-%20003.pdf.

Krisnamurthi, Bayu. (2003). The agenda for empowering farmers in the context of consolidating national food security. *Journal of People's Economy*, Year II, Number 7.

Koentjaraningrat. (1984). A model for studying social and cultural factors that influence diet and food consumption patterns. Issue: Bhratara Karya Script

Mewa, A. (2002). Food Security, Concept. *Measurement and Strategy. Agro-Economy Research Forum, 20*(1), 12. https://doi.org/10.21082/fae.v20n1.2002.12-24

Michael Quinn Patton. (1990). Qualitative evaluation and'research methods, second edition, Sage Publication.

Norge W. Jerome, Rnady F. Kandel, and Gretel H. Pelto (eds.), Redgrafe Publishing Company.

Novianti, K., Warsilah, H., & Wahyono, A. (2013). Perubahan iklim dan ketahanan pangan masyarakat pesisir (Climate change and food security on coastal community). *Jurnal PKS, 15*(3), 203–218.

Pakpahan, A.; Saliem, H. P., Suhartini dan S. H., Syafa'at, N. (1993). Research on food security for low-income communities. Monograph Series No. 14. Center for Agricultural Socioeconomic Research, Bogor.

Putnam, Robert. (1993). The prosperous community: social capital and public life. *The American Prospects, 13.*

Pollnac, Richard, B. (1988). Social and cultural characteristics in the development of small-scale fisheries. in the book prioritizing humans in development. *Sociological Variables in Rural Development (Michael M. Cernea)*. Jakarta: UI Press

Rustanti, N. (2016). *Textbook of Food and Nutrition Economics*. Deepublish.

Reutlinger, S. (1987). Food security and poverty in developing countries, In Gittinger J.et.al. (eds) Food Policy: Integrating Supply, Distribution and Consumption. UNICEF 2016, Humanitarian appeal report, Addis Ababa Ethopia

Sayogyo. (1972). Modernization without development in rural Java. *Bogor, Research Institute for Rural Sociology*, IPB.

Simatupang, P. (1999). Toward sustainable food security: the need for a new paradigm in simatupang, P. et al. (eds.) Indonesia's economic crisis: effects on agriculture and policy responses. *Centre for International Economic Studies*, University of Adelaide 5005 Australia

Suryahad Asep, Daniel Suryadarma; Sudarno Sumarto; Jack Molyneaux. (2006). Agricultural demand linkages and growth multiplier in rural Indonesia. *Working Paper, SMERU Research Institute.*

Soetrisno, N. (1995). World food security. *Concept, Measurement and Dominant Factors. Food Magazine V*(21).

Stern Nicholas. (2007). The approach to the *Review* in the Oxonia. Intergovernmental Panel of *Climate Change* (IPCC).

Stern, N. (2008). The economics of climate change. *American Economic Review, 98*(2), 1–37.

Suhardjo. (1989). *Pangan, gizi dan pertanian.* Penerbit Universitas Indonesia.

Suhardjo. (1996). *Berbagai cara pendidikan gizi.* Bumi Aksara.

Suhardjo, L., & Brady, and Judy. (1986). *Food, Nutrition and Agriculture.* UI Press.

Suryana Achmad. (2014). Toward Sustainable Indonesian food security 2025: challenges and its responses. https://media.neliti.com/media/publications/56153-ID-menuju-ketahanan-pangan-indonesia-berkel.pdf.

Suarni and Widowati, http://balitsereal.litbang.pertanian.go.id/wp-content/uploads/2016/11/tiganol.pdf).

Tjondronegoro, Sediono, M. P. (1999). Fading village autonomy (pp. 15–25) In: pieces of sociology from rural areas. *Directorate General of Higher Education, Ministry of Education and Culture of the Republic of Indonesia.* Jakarta.

Uphoff, Norman. (1986). Local Institutionaln development: an analytical sourcebook with cases. Kumarian Press.

Warsilah, H. (2011). Penerapan kebijakan ketahanan pangan bagi pencapaian kedaulatan pangan. Makalah disampaikan pada Konferensi Ilmiah Pangan Nasional (KIPNAS), Indonesia.

Warsilah Henny. (2012). Qou vadis local food sovereignty: restoring local food sovereignty. News Source http://pangan.banglikab.go.id/

Warsilah Henny, (2013a). The role of food habits in coastal rural communities in supporting food security: the case of bahoi and bulutui villages, north minahasa. *Journal of Society & Culture,* **15**(1)

Warsilah Henny. (2013b). Environmental damage impacts community food security. https://nationalgeographic.grid.id/read/13286818/rusak-environmental-berdampak-pada-ketahanan-pangan-community

Warsilah Henny. (2013c). Rusaknya lingkungan berdampak pada ketahanan pangan masyarakat. https://nationalgeographic.grid.id/read/13286818/rusaknya-lingkungan-berdampak-pada-ketahanan-pangan-masyarakat.

Wahyono Ary. (2011a). LIPI recommends the form of a non-rice food bank, https://kominfo.jatimprov.go.id/read/umum/26274;

Wahyono Ary. (2011b). LIPI rekomendasikan bentuk bank pangan non beras https://kominfo.jatimprov.go.id/read/umum/26274.

Self-Sustaining Community Resilience Through Local Asset Management: A Case Study of the Local Community Association 'Shinjyo Aigoukai' in Shinjyo Town of Tanabe City, Wakayama, Japan

Miki Yoshizumi

Abstract The Great East Japan Earthquake hit Tohoku in 2011, causing extensive damage. Researchers and governments forecasted another large earthquake and tsunami to hit the Wakayama and Toukai areas. Through sustainable local asset management, a local community association known as 'Shinjyo Aigoukai' implemented several activities to address tsunami disasters without relying on local government support. Shinjyo Junior High has developed tsunami disaster education programs with support from the association, working together to learn lessons from past tsunami events. This chapter assesses self-sustaining community resilience through local community and asset management using natural resources by observing 'Shinjyo Aigoukai' in Tanabe City.

Keywords Community resilience · Local asset management · Tsunami disaster preparedness · Grassroots initiatives · Disaster education · School relocation

1 Introduction

1.1 Natural Disasters in Asia

As the world enters the second millennia, the frequency and magnitude of natural disasters seem to have increased significantly. Nowadays, it is not too much of an exaggeration to say that virtually every week, we hear from mass media about the occurrence of natural disasters and their devastating damage in many parts of the world. Large-scale disasters have significant humanitarian, social, security, political and economic implications.

M. Yoshizumi (✉)
College of Gastronomy Management, Ritsumeikan University, Kusatsu, Japan
e-mail: miki-yo@fc.ritsumei.ac.jp

M. Yoshizumi et al. (eds.), *Navigating Local Sustainability in Food, Community, and Innovation*, https://doi.org/10.1007/978-981-96-8963-7_4

Records show that the frequency of natural disasters have increased, particularly since the 1990s. Among the many implications, its economic impact has increased significantly. However, the degree of this impact varies from country to country, especially between developed and developing countries, as the per capita cost of disasters relative to GDP is much higher in developing countries. It has been estimated, however, that worldwide economic losses due to natural disasters totaled US$162.3 billion in 2013, which rose to over US$270 billion in 2022 (Munich Re, 2024). The number of natural disasters that have occurred over the past ten years has more than doubled since the 60s.

Furthermore, other dimensions, such as social and environmental factors, are equally significant as the economic impacts in vulnerable communities. Such disasters contribute to the fragmentation of communities and societies, including the loss of social identity, breakdown of social norms and loss of informal social safety nets.

Disaster management is intrinsically linked to sound environmental management, which encompasses many dimensions of productivity in the daily lives of millions of people, particularly the poor, whose lives and livelihood, such as agriculture, fishery and forestry, are dependent on the natural environment and its resources. In such vulnerable communities, what matters is a bottom-up management approach, where groups of communities by themselves can become instrumental in making important decisions pertaining to their surrounding environment and natural resources.

Natural disasters and disaster management are crucial in all regions of the world, particularly Asia. It is quite remarkable that in Asia alone, we have witnessed the following:

- Thirty-eight percent of all natural disasters worldwide;
- Ninety-one percent of all people killed worldwide in 2005;
- Eighty-three percent of the total number of affected people worldwide;
- Ninety-one percent of all human losses between 1975 and 2005

(The Asian Disaster Reduction Center, 2005).

It is obvious that Asia's high vulnerability is exemplified by the numbers above. Among the types of natural disasters and environmental issues, Asia has virtually all types of disasters, including earthquakes, floods, cyclones, volcanos, droughts, extreme temperatures and environmental degradation problems. The two most devastating natural disasters in recent times—the mighty tsunami on 26 December 2004 and the massive earthquake on 8 October 2005—have left many communities and lives in Asia devastated.

Certain types of natural disasters, such as hydrometeorological disasters, can be mitigated and even prevented with cautious environmental management. However, it has been reported that, of the natural disasters that have occurred in Asia, as much as eighty percent are hydro-meteorological or climate-related, further underlining the importance of disaster management, with the integration of sound environmental management in the region (ADPC 2004).

1.2 Community-Centered Approach

Human security is a key component of disaster management, with an apparent link between disaster and on-the-field environmental management. To achieve human security, sustainable and effective disaster management must be implemented using a simple yet effective community-based approach.

The importance of the community and its involvement has been recognized and practiced throughout the academic realm, as well as the international community. It suffices to say that in 2005, the Hyogo Framework for Action has recognized the community-based approach as an 'indispensable component for disaster management.' Previously, however, disaster management, conflict prevention, and capacity-building strategies used to concern 'mainly official authorities, and not the communities and their leaders who are at risk' (Commission on Human Security, 2003). Therefore, a community-centered approach involving as many people as possible is essential to complement the institution-driven justice and reconciliation processes.

It is necessary to empower people and communities in vulnerable regions. It was previously assumed that when a natural disaster occurs, it is the role of the government or national authority to tackle or mitigate it. However, as seen in every natural disaster event, local vulnerable communities are impacted the most. Moreover, while occasional large catastrophes, such as earthquakes, volcanic eruptions, and cyclones, continue to occur in various parts of the world, the rapid increase in disaster loss is almost entirely due to an exponential increase in the occurrence of small-to medium-scale disasters associated with socio-natural hazards, such as landslides, floods, droughts, and fires (Yodmani, 2001).

With such changes in disaster patterns, the community-based approach offers one of the few alternatives for managing and reducing risks in vulnerable regions of developing countries.

The Great East Japan Earthquake hit Tohoku in 2011. Given the extensive damage, researchers and governments have eanalysed and forecasted that another large earthquake and tsunami will hit the Wakayama and Toukai areas. Faced with such tsunami disaster risks, local governments in Japan require the development of urgent disaster management and community resilience. Shinjyo Town is located in Tanabe City, Wakayama Prefecture, which is one of the areas with the highest risk of tsunamis. Tsunamis strike Shinjyo every 100–200 years. To address tsunami disasters and develop community resilience, a local community association known as 'Shinjyo Aigoukai' has supported local communities and schools through asset management using local natural resources. Through sustainable local asset management, the association has implemented several activities to address tsunami disasters without relying on local governments. Shinjyo Junior High has developed tsunami disaster education programs with support from the association, and they have worked together to learn lessons from past tsunami events and gain knowledge on tsunami survival.

This chapter examines self-sustaining community resilience through local community and asset management using natural resources, illustrating a case study of 'Shinjyo Aigoukai' in Tanabe City, Wakayama Prefecture, which does not rely on government support.

2 Community Background

2.1 *Significance of the Community, History, Culture and Geography*

Tanabe City is located in the coastal areas southwest of Wakayama Prefecture. Tanabe is a quake-prone area. Massive earthquakes and tsunamis have hit the coastal areas of the Wakayama Prefecture every 100–200 years. In 1946, an earthquake with a magnitude of 8 occurred in the Wakayama area, killing 269 people. Tanabe City experienced some of the most severe damage. In 1946, 69 people were killed in Tanabe City, which had the highest number of casualties in the Wakayama Prefecture. At present, it is forecasted that such a large earthquake will occur within approximately 30 years (Ministry of Land, Infrastructure, Transport and Tourism, 2014). According to surveys by the Japanese national government, it is forecasted that a 12 m tsunami could hit the coastal areas of Tanabe City.

The Shinjo area is located in Tanabe City and is one of its most vulnerable areas. To reduce the damage from earthquakes and tsunami disasters, locals in the Shinjyo have addressed disaster prevention by themselves, not relying on the local government (Fig. 1).

2.2 *Evolution of Community Issues*

Shinjyo Village was integrated into Tanabe City in 1954. It contains owned lands and forests of village property as the common land of villagers since the Edo era. When Shinjyo was integrated, Shijyo Village contributed forty percent of the land common to Tanabe City. However, it retained the remaining sixty percent of the common land of the local Shinjyo people.

First, the Shinjyo Agriculture Association managed common land. To use it effectively for the education and safety of local people, an incorporated association for the community, the Shinjyo Aigoukai Association, was established in 1964 (Shinjyo Aigoukai, 2015).

Fig. 1 Location of Tanabe City and Shinjyo Area

3 Introduction of the Community-Based Initiative

3.1 Purpose of the Initiative

The association managed common land and worked for education, welfare, and disaster management in the Shinjyo area. The Shinjyo Aigoukai Association supports various local community activities. Figure 2 shows a breakdown of expenses for each activity of the association. Since it was established, the fund which was derived from the revenue generated by the utilization of land that was once common forestland during the time Shinjyo was a village has been used for educational and social activities of the local Shinjyo people.

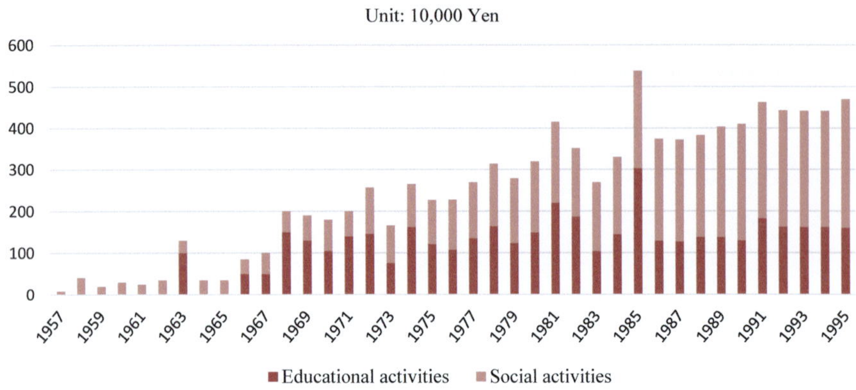

Fig. 2 Expenses of educational and social activities of Shinjyo Aigoukai Association

3.2 *Explanation of the Origin, Evolution and Current Status of the Initiative, with Descriptions of Key Stakeholders/individuals*

The Shinjyo Aigoukai Association has supported and developed various community activities and projects for local resilience using its own funds and not relying on government support. To continue their independent activities, they attempted to manage their assets sustainably. Figure 3 shows the types and transitions of income of the Shinjyo Aigoukai Association. Until 1967, the main income was from the sale of matsutake mushrooms. Since 1968, the Shinjyo Aigoukai Association has lent part of the common land for a golf course and the main income of the association has been from the lease. Furthermore, since 1975, it has lent part of the common land for parking. The income from the golf course lease and parking contributes to stable revenue for Shijyo Aigoukai. However, matsutake mushrooms were difficult to find because of the decline of pine trees. Additionally, before timber from the common forest could be sold, the sale price of local timber was low; therefore, the cost of maintaining the forest was much higher than the sale price.

Invested capital was used for local community, educational, cultural, and welfare activities. The Shinjyo Aigoukai Association attempted not to reduce its assets. Figure 4 shows the balance between revenue and expenditure from 2009 to 2013. In 2009 and 2010, revenue was higher than expenditure. In 2011, they used their funds to build a gymnasium for the Shinjyo Elementary School. In 2012, however, the Shinjyo Aigoukai Association restrained spending and maintained a balance. Figure 5 shows the floating and fixed assets. Assets are sustained by balancing revenue and expenditure.

Fig. 3 Shifts in Shinjyo community Hometown Association's income

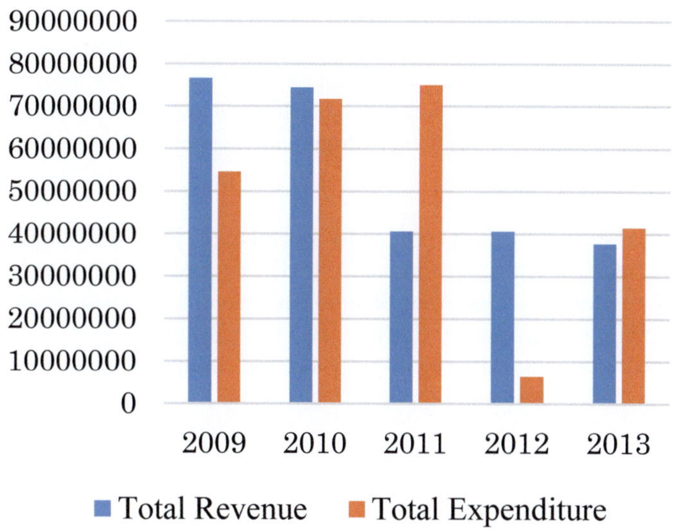

Fig. 4 Balance between revenue and expenditure of Shijyo Aigoukai Association fund

4 Description of Initiative's Activities

The types of activities of the initiative are as follows:

(1) School education

The association provides grants to nursery schools, kindergartens, elementary schools, junior high, and high schools in the Shinjyo area to improve their school buildings, educational environments, and purchasing of books.

(2) Social education

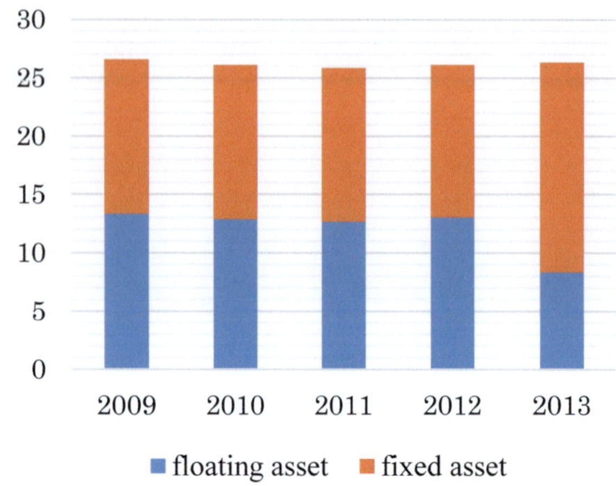

Fig.5 Floating and fixed assets of Shinjyo Aigoukai association

The association provides grants to each eighbourhood association and community center in the Shinjyo area. Grants have been used to improve buildings, community environments, and community events. The Shinjyo Aigoukai Association covered the cost of renovating the main Shinjyo community center in 1985.

(3) Sustaining local culture

There are local shrines for each community as well as the local main Ogata shrine in the Shinjyo area. The Shinjyo Aigoukai Association provides grants to each shrine for building renovations. Moreover, Shinjyo has several local festivals, dances, and songs. The Shinjyo Aigoukai Association also provides grants to local community groups to preserve their cultures. In addition, the Shinjyo Aigoukai Association provides grants to a local study group trying to publish a local history book on the Shijyo area.

5 Disaster Management

5.1 Developing an Evacuation Space

In March 2008, Shijyo Aigoukai constructed an evacuation space for tsunamis using common land (Fig. 6).

Fig. 6 Photo of the new evacuation space. *Source* Shinjyo Aigoukai Association HP

5.2 *Building Watermarks of Historical Tsunamis*

The Shinjyo Aigoukai Association built watermarks of historical tsunamis in 1854 and 1946 at the areas of frequent tsunamis (Fig. 7).

5.3 *Relocating Schools to Higher Elevations*

Previously, Shinjyo Elementary School and Shinjyo Junior High School were located near the coastal area at low altitudes, where they were in danger of tsunamis. Locals asked the Tanabe City government to relocate the schools to higher altitudes, but as the cost was too high, the request was not realized. To reduce potential risks to students, the Shinjyo Aigoukai Association financially supported schools to move to higher altitudes. Shinjyo Elementary School was relocated to a higher altitude in 1967 and Shijyo Junior High School was relocated in 1983. The Shinjo Aigoukai Association paid 21,930,000 yen (approximately 205,337 USD as of February 2018) for elementary schools and 82,190,000 yen (approximately 769,570 USD as of February 2018) for its relocation projects. The Shinjyo Aigoukai Association not only provided financial support but also found land for relocation projects. These schools are designated as evacuation centers in Tanabe City during tsunamis.

Fig. 7 Tsunami watermarks in Shinjyo Area Left: Watermark of the 1946 Tsunami, located at Uchinoura Park. Right: Watermark of the 1854 Tsunami, located in Yamagi Shrine

5.4 Promoting Disaster Education Programs

Shinjyo Aigoukai Association published a book, 'Tsunami of Showa era', to teach people about tsunami risks, actual tsunami experiences, and lessons, with many photos of the tsunami and the damage that occurred in 1946 (Fig. 8).

Shinjyo Junior High School developed 'Shijyo Jishingaku (the study on earthquake and tsunami in Shinjyo)' in 2001, in partnership with the Shinjyo Aigoukai Association, using the class 'hours for comprehensive studies', which was promoted by the Japanese Ministry of Education, Culture, Sports, Science and Technology from 2000. They developed seven tsunami disaster education programs based on existing subjects, such as mathematics, English, and music (Shinjyo Junior High School, 2013). In 2014, the main theme was 'connecting with local communities and connecting from the past to the future'. In particular, they attempted to learn about the local history and culture related to tsunamis. Teachers educated students on how a tsunami is not a mere natural disaster but part of the local environment. To learn about local cultures, students conducted field surveys of local communities and the environment and conducted interviews with locals who experienced the tsunami in 1946. Based on the interviews, they created hazard maps that showed which areas suffered severe damage from the tsunami and which areas survivors evacuated to. Subsequently, the students presented the lessons that they learned and their research results to the locals. Additionally, the school publishes a calendar and report for

Fig. 8 Relocating schools to higher elevation in Shinjyo Created by the author using Google Maps

locals, which provides research results and the necessary information about tsunami evacuation.

Moreover, the school organizes emergency training for the locals. During the training, they time how long it takes people to evacuate without advance notice. Furthermore, the junior high school provides lectures for local nursery schools, primary schools, and elderly facilities; the junior high school students give them educational lectures. Through the lectures, students can learn about tsunami disasters in depth and can take the initiative in tsunami disaster management.

6 Outcome of Activities

These activities have several outcomes. First, the facilities of some schools in Shinjyo have been improved. For example, a new school building, Shinjyo Elementary School, was built and funded by the Tohoku Aigoukai Association. The building is made of valuable wooden architecture that uses local materials, which are much more expensive than concrete and imported materials. Additionally, various school items and books for school libraries were donated.

Second, facilities and items for local community activities were donated to sustain local activities. Traditional festivals have continued with the support of the Shinjyo Aigoukai Association.

The third output is for disaster management. The Shinjyo Aigoukai Association has developed publications and watermarks of historical tsunamis to raise awareness about natural disasters. Moreover, the Shinjyo Aigoukai Association has developed evacuation spaces during tsunamis and relocated schools to higher ground. 'Shijyo Jishingaku', the disaster education program working with Shinjyo junior high school, has continued for over 10 years. Shijyo Jishingaku was awarded the Japanese Disaster Education Award (Bousai Koushien) in January 2015 (Disaster Prevention Research Institute, 2015).

7 Conclusions

Community-based disaster management is necessary. However, many disaster management projects that are promoted by the government and locals are often neglected. In Japan, when local governments organize evacuation training, only a few elderly people participate.

The case study of the Shinjyo Aigoukai Association provides one of the best practices for initiating and motivating locals, not only for disaster management but also for local community activities. In particular, Shinjyo Aigoukai's asset management provides methods for local community initiatives that do not rely on government support.

References

Asian Disaster Reduction Center. (2005). *Natural Disasters Data Book*. Asian Disaster Reduction Center, Kobe.

Center, A. D. P. (2004). *Environmental degradation and disaster risk*. Asian Disaster Preparedness Center, Bangkok.

Commission on Human Security, Human Security Now. US: New York; (2003).

Disaster Prevention Research Institute, Kyoto University. Disaster education challenge plan (http://bosai-study.net/houkoku/plan16/index.html) Accessed 10 July 2015.

Ministry of Land, Infrastructure, Transport and Tourism. (2014). Map showing areas with the potential for flooding from tsunami (http://www.mlit.go.jp/river/shinngikai_blog/shaseishin/kasenbunkakai/bunkakai/dai50kai/siryou3-13.pdf). Accessed 28 April 2015.

Munich Re, Topics Geo. Annual review: Natural catastrophes. (https://www.munichre.com/en/risks/natural-disasters.html). Accessed 10 October 2024.

Shinjyo Aigoukai (http://www.aikis.or.jp/~shinjyo-aigokai/index.html). Accessed 21 June 2015.

Shinjyo Junior High School. (2013). Shinjyo Jisingaku Final Report.

Yodmani, S. (2001). *Disaster risk management and vulnerability reduction: Protecting the poor*. Asian Disaster Preparedness Center, Bangkok.

Building Up Resilience for Small-Medium Enterprises in Traditional Food Villages: A Case Study of Central Vietnam

Thi Kinh Kieu, Thuy Xuan Cam Vo, Aurora Lokita, and Nini Purwajati

Abstract Traditional small and medium food enterprises (SMEs) significantly improve local economies, raise incomes, end hunger and poverty, generate employment, attract idle capital from the population, and enhance the flow of capital throughout society. Particularly, during the COVID-19 pandemic, traditional food SMEs helped ensure food security in many cities where supply chains were disrupted and delayed. However, traditional food SMEs face various challenges, such as limited access to modern technology and markets, food fraud, undervaluation, and external shocks. This chapter details the existing vulnerabilities of a traditional fish sauce village in Da Nang City, Central Vietnam, and the different efforts of the SMEs in this village to build their adaptability and resilience capabilities.

Keywords Fish sauce · Nam O · Resilience · SMEs · Da Nang City

1 Introduction

Small and medium-sized enterprises (SMEs) are companies that fall below a specific threshold in terms of revenue, assets, or workforce size (Berisha, 2015). Small and medium-sized businesses, particularly in transitional countries, are important engines of economic growth (Obi et al., 2018). SMEs account for 99% of all enterprises in the EU, provide jobs for an estimated 100 million people, and

T. K. Kieu (✉)
The University of Danang, University of Science and Education, Danang, Vietnam
e-mail: ktkinh@ued.udn.vn

T. X. C. Vo
The Center of Building Up Sustainability, Danang, Vietnam

A. Lokita · N. Purwajati
Resilients Cities Network, Singapore, Singapore
e-mail: alokita@resilientcitiesnetwork.org

N. Purwajati
e-mail: npurwajati@resilientcitiesnetwork.org

M. Yoshizumi et al. (eds.), *Navigating Local Sustainability in Food, Community, and Innovation*, https://doi.org/10.1007/978-981-96-8963-7_10

provide more than half of the GDP of the EU (https://single-market-economy.ec.europa.eu/smes_en). Global economies benefit greatly from the contributions of SMEs: 70% of all jobs and 99% of businesses in OECD nations are accounted for by SMEs; in high-income nations, SMEs also make up more than 50% of the GDP (https://www.mckinsey.com/industries/public-sector/our-insights/beyond-financials-helping-small-and-medium-size-enterprises-thrive). SMEs are essential for the development of a nation's economy because they drive competition in terms of efficiency, cost, and product design.

However, SMEs have been severely affected by the COVID-19 pandemic. A large proportion (70–80%) of SMEs in 30–32 nations lost 30–50% of their income between February 2020 and April 2021 (https://www.mckinsey.com/industries/public-sector/our-insights/beyond-financials-helping-small-and-medium-size-enterprises-thrive). It is thought that SMEs continue to grapple with financial strain, supply chain disruption, hiring issues, adapting to new environmental conditions, and vulnerability and resilience (https://www.iifl.com/blogs/business-loan/5-challenges-for-small-businesses-post-covid-19; Erdiaw-Kwasie et al., 2023; Naradda Gamage et al., 2020). The beneficial effects of the innovative activities of SMEs on their survival and performance are strengthened when they obtain external support through funding, knowledge transfer, and training (Adam & Alarifi, 2021). For example, during the COVID-19 pandemic, governments provided tax incentives, state guarantees for loans, and direct financial relief (https://www.mckinsey.com/industries/public-sector/our-insights/beyond-financials-helping-small-and-medium-size-enterprises-thrive). To maintain their viability, SMEs must overcome several obstacles.

In Vietnam, SMEs account for 36% of the country's value addition and 47% of all employment (OECD 2021). The Vietnamese government has implemented several policies to support SMEs. The SME Support Law enacted in 2018 demonstrates the government's commitment to promoting domestic SMEs. Similar to other countries, the COVID-19 pandemic has significantly influenced SMEs in Vietnam, particularly in the food processing industry involved in tourism. Studies by Ninh Thuan (Wu et al., 2022) and Thanh Hoa and Tuyen Quang (Nguyen & Huong, 2021) indicate a remarkable drop in sales and employment as a result of COVID-19. Some studies have emphasised the importance of effective and innovative policies, changes in corporate governance principles, and the role of government support in helping SMEs overcome the adverse effects of the pandemic in Vietnam (Le & Nguyen, 2022; Ngo, 2023). Major changes in the business activities of SMEs have been brought about by post-Covid, however, research on this topic is scarce in Vietnam. Therefore, this chapter will contribute to both practical and academic knowledge by detailing post-Covid adaptation actions in traditional fish sauce craft villages.

2 The Nam O Fish Sauce Village

The Nam O fish sauce village area was formed in the early twentieth century and is located approximately 3 km from the foot of the Hai Van Pass to the south of the city. The study site is located in the territory of two wards, Hoa Hiep Bac Ward and Hoa Hiep Nam Ward, both of which belong to the Lien Chieu District, Da Nang City (Fig. 1).

In terms of administrative boundaries, there are three main residential areas in the village: (i) Nam O 2.1, (ii) Nam O 2.2, and (iii) Xuan Duong, with a total area of approximately 3.65 km^2, and more than 5800 people in over 1000 households.

Approximately 93 fish sauce-making households actively operate in Nam O village in the Hoa Hiep Bac and Hoa Hiep Nam wards. Among these households, 64 are members of the Nam O Fish Sauce Traditional Craft Village Association, with 27 small- and medium-sized establishments having registered their own brands, two cooperatives and one enterprise (Table 1 and Fig. 2).

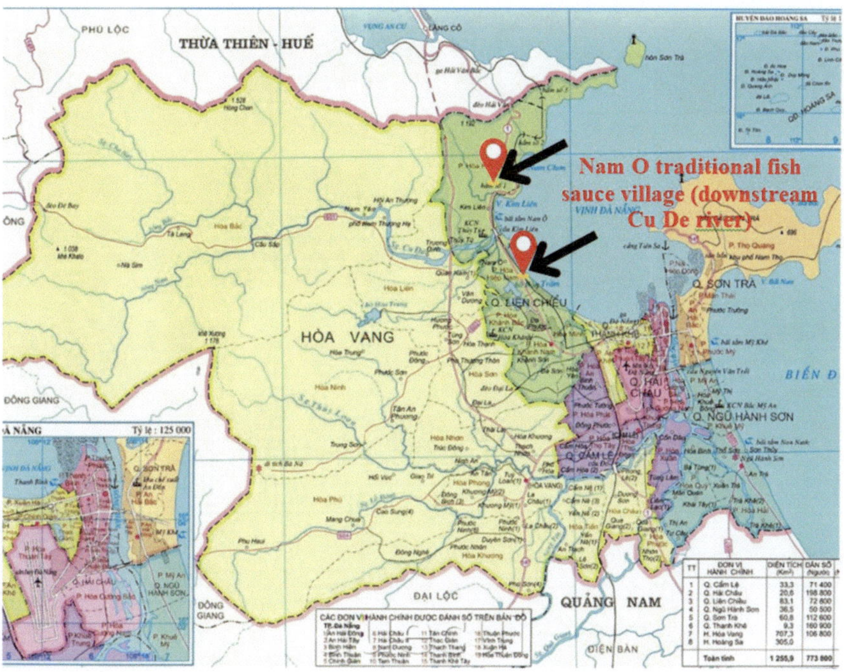

Fig. 1 Geographical map of Nam O fish sauce village. *Source* official map of Da Nang City and edited by the authors

Table 1 List of SMEs in Nam O village

SME categories / Year	Household[a]	Small and registered[b]	Medium (cooperatives/collective)[1]
2019	35	27	4 Cooperatives (O Long Cooperative, Dong Hai Cooperative, Xuan Thieu Cooperative Bình Minh Cooperative)
2020	35	27	2 Cooperatives (O Long Cooperative, Bình Minh Cooperative) 1 Enterprise (Hong Huong Fish Sauce Co., Ltd)
2021–2022	35	27	2 Cooperatives (O Long Cooperative, Bình Minh Cooperative) 1 Enterprise (Hong Huong Fish Sauce Co., Ltd)

[a]For household businesses, staffed by 1–2 people, fish sauce production ranges from 3040 L/month
[b]For small and registered, staffed by 2–4 people, fish sauce production ranges from 100 L/month

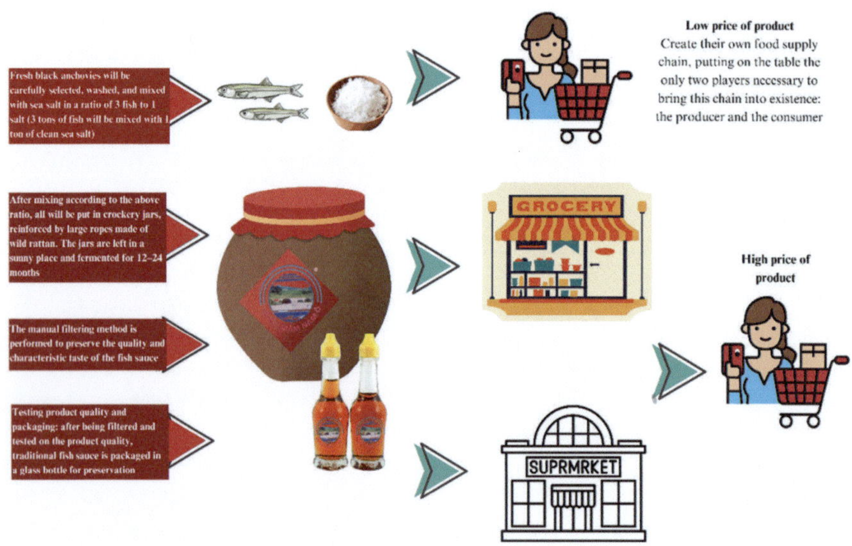

Fig. 2 Fish sauce productiofn overview

3 Challenges, Shocks and Stress

Based on the results of in-depth interviews and secondary data, the shocks and stresses related to livelihoods in the community and SMEs in Nam O village are as follows.

[1] For medium (corporations/collectives), staffed by 7–10 members, the average production is about 500–1000 L/month.

3.1 Acute Shocks

- **Climate change**: Climate change is one of the reasons for the increase in tropical monsoon climate so Nam O is directly affected by the strong intensity of natural disasters (including inundation, storms, and tropical depressions) over time. Being close to the sea, Nam O village was affected by the high volume and frequency of rainfall, which negatively affected the quantity and quality of raw materials and the quality of fish sauce produced in Nam O.
- **The COVID-19 pandemic**: For cooperatives and businesses, sales decreased by half as shopping malls and supermarkets were temporarily closed owing to social distancing regulations. However, during the pandemic, household fish sauce production significantly increased product consumption, growing to 1.5–2 times higher than the output before the pandemic (for small and medium businesses). This might be explained by the fact that during the pandemic, the source of goods from other localities was interrupted by the city's social distancing regulations, whereas the number of local customers from personal relationships was higher. Therefore, the residents of Da Nang tended to consume local products, specifically Nam O fish sauce. The results of the in-depth interviews showed that fish sauce production decreased again after the pandemic, to the same level as pre-epidemic consumption.

3.2 Chronic Stresses

- **Disappearance of Fishery**: According to the city's plan in the "Nam O Ecotourism Resort" project, the anchorage and sheltering places for boats were moved away from the fishermen's living area in order to make way for project implementation. This is considered one of the reasons why fishing jobs have faded away in Nam O village based on the fishermen's traditional view that "Assets are attached to people". In addition, inshore fishing activities by high-powered vessels are strictly prohibited[2]; as a result, the number of households changing from fisheries to other occupations (specifically manual workers and freelance workers) has increased. To encourage people to conduct offshore fishing, the People's Committee of Da Nang City has directed the Department of Industry and Trade to support social capital loans for fishermen for the purpose of purchasing tools for offshore fishing activities. However, based on the results of in-depth interviews, it was noted that local fishermen do not have sufficient capacity and experience to transition from inshore fishing to offshore fishing; therefore, borrowing money is not a viable option due to the high financial risks involved.

---

[2] Decision No. 8329/QD-UBND dated the 19th of October 2007 on the Program to protect and develop Da Nang's fisheries resources by 2010 with a vision to ensure seafood resource security in the city by 2020.

- **The quantity of input materials**: The quality and quantity of input materials for traditional fish sauce-making are considered one of the livelihood risks for the local community. Anchovies used in fish sauce production are primarily sourced from Sa Huynh (Quang Ngai Province). Based on the results of in-depth interviews, the quality of fish from Sa Huynh is unsatisfactory because it is frozen (mixed with fresh water), which does not meet the requirements of traditional fish sauce production in Nam O. Consequently, the quality of the fish sauce is not as good as during the period before 2010. The process of traditional fish sauce production has not been standardised or published, leading to the management of food hygiene and safety not being strictly implemented and instability in the quality of Nam O fish sauce products.
- **Lack of standardized process of making fish sauce**: Many establishments do not make fish sauce following traditional recipes. For example, the fermentation process lasts for more than 12 months; however, some businesses shorten this time (10 months). This leads to variable product quality, which affects brand perception in the market.
- **Cost of traditional fish sauce production**: The increase in production capital puts pressure on households to produce fish sauce freely. One possible reason for this is the increasing costs of input materials and fish sauce containers due to inflation (for instance, the cost of high-quality salt has increased from 100,000 VND/4.26 USD/50 kg in 2020 to 200,000 VND/8.52 USD[3]/50 kg in 2022). Notably, the source of output for consumers is not guaranteed, leading to a lack of stability in the market price of the Nam O fish sauce to ensure capital recovery. Therefore, the proportion of households producing traditional fish sauce tended to level off and gradually decrease.
- **Lack of SME capacity in developing the market**: For enterprises and cooperatives, with the advantages of accessing digital technology and having a team of experts in market development, the average production is approximately 500–1,000 L/month. For licenced production establishments, the output of fish sauce in the domestic market is approximately 100 L/month. The customers of these establishments were health-oriented supermarkets and mass consumers. However, for production households and freelance business households, the production of fish sauce ranges from to 30–40 L/month, and the customer source depends on personal relationships formed by the businesses. Based on the customer sources of the three business models, it is evident that more than 50% of fish sauce-producing households lack the capacity to promote and develop the market for their fish sauce products, resulting in unstable livelihoods. According to the results of the survey, small- and medium-sized households/production establishments need to shape their individual brands, specifically regarding product designs and stable customer development strategies in the domestic fish sauce market, under the management of the Nam O Fish Sauce Traditional Association. Furthermore, very few young people in the village are interested in or desire to work in the fish

[3] Conversion based on 1 USD = 23,617 VND (https://www.xe.com/currencyconverter/convert/?Amount=1&From=VND&To=USD, retrieved on Feb 6, 2023).

sauce production industry; only three business owners are under 40 years old, and the majority of the remaining proprietors are over 65 years old.

- **Higher price and organic taste of Nam O Fish Sauce**: Some customers who have used Nam O fish sauce state that there are two reasons for this: (i) the average price of fish sauce in the market is higher than that of industrial fish sauce products, and (ii) the taste of salty traditional fish sauce and its strong smell are not popular among ordinary consumers.
- **The use of plastic bottles**: Use of plastic bottles to contain and preserve fish sauce products decreases production costs. Thus, plastic bottles are a priority for dealing with the problem of capital. Similarly, plastic bottles decrease the costs and risks of delivery. However, product preservation in plastic bottles leads to an increase in chemical hazards, which negatively affect human health.

4 Building up Resilience

4.1 External Support

SMEs in Nam O fish sauce village have received support from the government and many organisations, including the following:

- Financial support: To increase production, SMEs were given discounted or interest-free loans, and households received subsidies during the COVID-19 pandemic. However, only six households successfully leveraged loans to increase output following the COVID-19 pandemic due of restricted land areas and production capabilities.
- Support for promoting the brand: The government assisted in the formation of the Nam O Fish Sauce Craft Village Association and has subsequently advised its members to register and declare the Fish Sauce Craft Village an intangible cultural resource at the federal level. The Nam O fish sauce production technique has become a geographical indicator product and is registered for intellectual property rights. Business homes were invited to participate in product promotions during the event.
- Encourage pride in craft villages and income diversification: Da Nang City's local education program includes craft villages, so some fish sauce artisans are welcomed by visiting groups to study first as students and pupils and then as tourists. It has been suggested that craft villages will become tourist attractions in future.
- Support research to improve product quality: The Department of Science and Technology of Da Nang City, Food Safety Department, universities, and research institutes in Da Nang have conducted research to ensure and improve the quality of Nam O fish sauce. Training sessions for SMEs on production hygiene, digital economy, and marketing have also been held.

- Support to improve capacity to respond to natural disasters and climate change: People in craft villages are able to participate in training on natural disaster risk reduction and forecasting the impacts of climate change on people in craft villages to proactively improve production and business activities.
- Help to enhance the quality of the environment: Previously, waste from fish sauce manufacturing was collected and used as animal feed. However, with the significant decline in the number of livestock households owing to urbanisation, this waste is now primarily discarded. mixed with residential garbage. Because of the volume produced, the waste collection unit frequently declines to collect this waste, which presents challenges for medium-sized production houses.

Evaluation of the effectiveness of external support activities

External assistance strengthens the ability of SMEs to conduct business, standardise the process of producing fish sauce, and increase their access to the consumer market. Medium-sized cooperatives can grow their companies and production with the aid of financial support. However, it is very difficult to develop the business scale in Nam O Fish Sauce Village because each household's land area is decreasing, and the village's business scale is limited. Both micro and small businesses believe that special loan packages are useless and that financial assistance is temporary.

Research and experiential learning initiatives, such as student study tours, are thought to be beneficial to Nam O's "pride of fish sauce making" in the long run. Because SMEs rarely invest in R&D, it is crucial to assist business households in their research endeavours to improve product quality and open new markets. To achieve efficacy, training initiatives must first evaluate the needs. Most senior SME owners find it extremely challenging to understand and implement marketing and digital economic concepts.

4.2 Internal Transformation

Online commerce and digital transformation are becoming unavoidable trends according to business households in Nam O Fish Sauce Craft Village. However, the application of these trends is not feasible for people aged >65 years old. The findings from three rounds of focus groups with SMEs indicate that business households want to create artisan villages together, even when they compete and do business together. The founding of the Nam O Fish Sauce Craft Village Association has helped showcase the resources of its members. The Association links business homes with external support units and the government and receives help in exchange for information.

> "The fish sauce business in Nam O has its own characteristics. Business households all have relatives or neighbors within the village, so reciprocity is very high. We are also aware that "big fish swallowing small fish" will lead to the disappearance of cultural and spiritual heritage values, reducing the brand of craft villages, and medium-sized business households

cannot compete with industrial production of fish sauce. Therefore, we must accompany each other and transfer our craft to interested younger generations to preserve the craft village" (Mr. Vinh, Head of the Fish Sauce Craft Village Association)

Stemming from practical needs, the "satellite" cooperative model, which is a combination of a medium-scale business household with small-scale business households to maximise efficiency and optimize each side's resources, was formed after the COVID-19 pandemic.

- Medium enterprise: The proprietor is a young individual raised in Nam O village. has a passion for protecting craft villages, works in the digital industry, promotes products using technology, and has an approach to production transformation (for instance, developing packaging for gifts or switching from a fish sauce bottle model to a glass bottle). However, production cannot be increased owing to the small production space. As a result, this ME has ties to nine small businesses that produce fish sauce before buying, bottling, and retailing.
- Small enterprises: The majority of SE owners are elderly and have previously made high-quality fish sauce (choosing fresh fish, salting, and incubating it, for example), but lack commercial expertise and sell their products at markets, mostly to acquaintances. Consequently, the selling price is low and profits are insubstantial (Fig. 3).

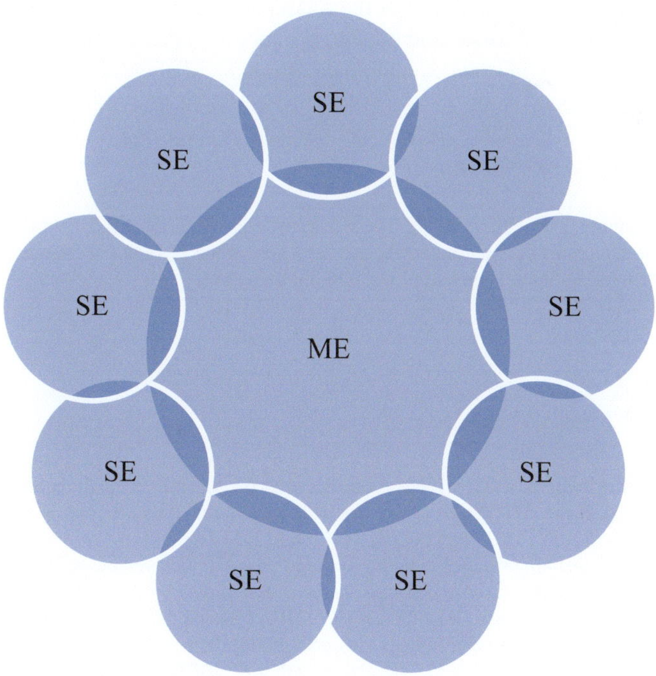

Fig. 3 Cooperation model of SMEs in Nam O

Although created just two years ago, this win–win strategy has already been proven successful. Each participant in the model has the opportunity to highlight their own abilities. The idea of openness in business, in which all participants are compensated for their contributions, had been discussed recently. It is hoped that the Craft Village Association will prove to be a successful business plan and draw on young, gifted individuals who will stay in the village to sustain the fish sauce industry in the years to come.

5 Conclusion

Modern business strategies and the rapidly changing digital economy pose many problems for traditional craft communities, both in Vietnam and worldwide. Most SMEs in craft villages create and distribute goods on a small scale, have manual production on a household scale, are run by older business owners, and lack R&D departments. Compared to industries that prioritise business optimisation, this results in greater production and running costs.

The preservation of Nam O Fish Sauce Village has benefited from analyses of the dangers and difficulties the village faces, as well as from solutions to assist with funding, marketing, and capacity-building. The establishment of the Craft Village Association strengthens the ability of the SMEs to withstand setbacks by providing a conduit between their voices and outside backers to guarantee the effectiveness of support. Furthermore, in order to guarantee efficient collaboration between companies in the fish sauce sector, the "satellite" cooperative model has also mobilised the resources of SMEs. This strategy can be used as a starting point to increase the resilience of SMEs in Vietnam.

Acknowledgements The authors would like to acknowledge the Global Resilient Cities Network for their grant to support the baseline study and activities in the Nam O Fish Sauce village. This book chapter would not have been completed without this grant.

References

Adam, N. A., & Alarifi, G. (2021). Innovation practices for survival of small and medium enterprises (SMEs) in the COVID-19 times: The role of external support. *Journal of Innovation Entreprenenurship, 10*, 15. https://doi.org/10.1186/s13731-021-00156-6

Berisha, G. (2015). Defining small and medium enterprises: A critical review. *Academic Journal of Business, Administration, Law and Social Sciences, 1*, 17–28.

Challenges For Small Businesses Post-COVID-19 | IIFL Finance. https://www.iifl.com/blogs/business-loan/5-challenges-for-small-businesses-post-covid-19

Erdiaw-Kwasie, M. O., Abunyewah, M., Yusif, S., & Arhin, P. (2023). Small and medium enterprises (SMEs) in a pandemic: A systematic review of pandemic risk impacts, coping strategies and resilience. *Heliyon, 9*, Article e20352. https://doi.org/10.1016/j.heliyon.2023.e20352

Helping small and medium-size enterprises thrive | McKinsey. https://www.mckinsey.com/industries/public-sector/our-insights/beyond-financials-helping-small-and-medium-size-enterprises-thrive

Le, T. T., & Nguyen, V. K. (2022). Effects of quick response to COVID-19 with change in corporate governance principles on SMEs' business continuity: Evidence in Vietnam. *CG*. https://doi.org/10.1108/CG-09-2021-0334

Naradda Gamage, S. K., Ekanayake, E., Abeyrathne, G., Prasanna, R., Jayasundara, J., & Rajapakshe, P. (2020). A review of global challenges and survival strategies of small and medium enterprises (SMEs). *Economies, 8*, 79. https://doi.org/10.3390/economies8040079

Ngo, Q.-H. (2023). The effectiveness of strategic alignment between open innovation and generic strategies: Empirical evidence from restaurant SMEs in Vietnam. *Journal of Open Innovation: Technology, Market, and Complexity. 9*, 100016. https://doi.org/10.1016/j.joitmc.2023.100016

Nguyen, T. L. H., Shah, F., Nguyen, H. A., Hoang, T. H. (2021). Impact of COVID-19 on small business households: A case study of Thanh Hoa and Tuyen Quang Provinces, Vietnam. *Nu Journal of Economics and Business, 1*, 1–11.

Obi, J., Ibidunni, A. S., Tolulope, A., Olokundun, M. A., Amaihian, A. B., Borishade, T. T., & Fred, P. (2018). Contribution of small and medium enterprises to economic development: Evidence from a transiting economy. *Data in Brief, 18*, 835–839. https://doi.org/10.1016/j.dib.2018.03.126

OECD. (2021). SME and Entrepreneurship Policy in Viet Nam. OECD.

SMEs-European Commission. https://single-market-economy.ec.europa.eu/smes_en

Wu, C. K., Nguyen, N. A., Dang, T. Q. T., & Nguyen, M.-U. (2022). The impact of COVID-19 on ethnic business households involved in tourism in Ninh Thuan, Vietnam. *Sustainability, 14*, 16800. https://doi.org/10.3390/su142416800

Sustainable Community Practices

Potential and Challenges for Community Energy in the United Kingdom: A Case Study Approach

Katalin Kuntner and Jane Singer

Abstract Small scale renewable energy projects run by communities may have various benefits for the community: the generated income can be re-invested locally, could lead to greater energy independence, support local jobs and strengthen community ties. It also tackles climate change on a local level as renewable energy is being used, while also makes people think about how to reduce their energy consumption. In the United Kingdom these projects have been supported by various subsidies and government policies leading to the emergence of more than 200 community energy organizations. However, with the substantial reduction of the government's feed-in tariff (FIT) programs these top-down incentives have disappeared in England, leaving future projects uncertain. Nevertheless, there had been many successful schemes, each unique on its own. This chapter will introduce four examples of community energy projects—micro hydro, wind turbines, solar power and attempts at a micro-grid—to illustrate the challenges and promise of these collaborations.

Keywords Community energy · Renewable energy transition · Local energy governance · Energy resilience · Grassroots energy initiatives · Policy and regulatory change

1 Defining and Explaining Community Energy

According to the UK government's definition, community energy is a collective initiative, with a strong emphasis on local engagement and leadership to reduce, purchase, manage and generate energy (Gov.uk, 2013). Community energy is commonly envisioned as a community generating energy for itself by investing in a renewable energy technology, but it may present itself in other forms as well. Some

K. Kuntner (✉)
Independent Researcher, Budapest, Hungary
e-mail: katalin.kuntner@gmail.com

J. Singer
Department of Global Studies, Kyoto University of Foreign Studies, Kyoto, Japan
e-mail: j_singer@kufs.ac.jp

M. Yoshizumi et al. (eds.), *Navigating Local Sustainability in Food, Community, and Innovation*, https://doi.org/10.1007/978-981-96-8963-7_5

communities might decide to invest in energy reduction technologies, such as insulation and draft proofing, instead of generation, or in trying to facilitate change in individual consumption patterns, or by implementing several activities at the same time. A community can be present in various ways in an energy project. The most common case is when people live in the same geographical area, thus forming a physical community (Cowtan, 2017). However, it is not always possible to initiate a project with only local people involved. If the community is too small, or not everyone is interested, share offers (opportunities to invest in the project) might open up to all potential investors, even those living on the other side of the country.

Community energy projects are run by an organization, typically a co-operative (co-op) or a community benefit society (BenCom). Both are forms of registered societies, but while co-ops are usually focusing on benefiting their members only, BenComs aim to benefit the community as well by supporting other projects with the generated income. This could include sporting facilities, new roofs for schools and other community facilities, planting trees, and changing conventional light bulbs to LEDs.

Community energy projects usually start with a few individuals taking interest in heat and/or electricity generation and forming a group to discuss ideas and possibilities, but as they don't necessarily have the expertise they might turn to other organizations and/or consulting companies for help. Looking for a suitable location by themselves is possible but it is also very time intensive, while consulting companies may already have the necessary information. Projects that failed to progress, e.g. because the commercial return was insufficient but where the land owner was interested, could be excellent candidates. For the next step, a feasibility study needs to be done which requires expertise, therefore they will most likely need to pay for a consultancy to do this for them. The study will reveal the possibilities of installing renewable energy technologies in the chosen location. According to one of the directors of a successful micro hydro project, one of the most important reasons to why some projects fail is that while they secure enough money to do the feasibility study, they don't have any left for the implementation of the project itself. If this is not the case, planning may start. This includes securing all the various planning permissions that are needed for such investment, including permits from the Environment Agency, landowners, etc. If all permissions are secured the project directors will open the share offer, which among many things includes information on the expected return and when and how shareholders will receive it. Most projects prefer to include community members, but if the necessary amount of money is not secured, they may choose to open the share offer to others as well. Regardless of the money invested, shareholders have one vote each.

The directors of co-ops and BenComs are usually residents of the community and their main job is to secure the project's development and operation and to deal with ordinary operational issues. In some cases, they might receive a small compensation, but they are often unpaid; the project is sustained by their sheer enthusiasm and commitment. The directors, who also invest, answer to the shareholders and have the responsibility to inform them about their activities and the progress of the project. Depending on the work that needs to be done, some projects also enlist the help of volunteers for everyday operations. As community projects there is an assumption of active participation from shareholders, but of course this varies depending on the individual. Nevertheless, as a democratic organization, people are always given the opportunity to voice their opinion and have the chance to offer assistance in developing and/or running the schemes. According to Cowtan (2017, p. 8), "One of the characteristics of community energy is that projects are being done *by* the community, not being done *to* the community".

This chapter, which is based on research conducted during a three-month internship at a non-profit organization in northern England, Cumbria Action for Sustainability, will present the current promise and challenges facing community energy in the United Kingdom by examining four specific community energy projects. The case studies were conducted based on site visits, participant observation, desk review and key informant interviews.

2 Why Does Community Energy Matter?

Humanity faces many challenges, including satisfying our growing energy needs with a system that mainly utilizes non-renewable energy sources that harm the planet and face eventual depletion. In order to create a more sustainable path for mankind we need to promote alternatives for our current energy system. Increasing attention is being paid to the decentralization and diversification of energy generation by utilizing locally available renewable energy sources. Decentralized energy production is a great way for a country to increase its resilience in case of power outages, and wide-scale utilization of renewables help decrease its dependency on imported fuels and mitigate the effects of climate change.

Renewable energy sources are available everywhere on Earth, much more so than fossil fuels, however they are not homogenously distributed. While wind and solar power can be found everywhere, their potential varies based on geographical characteristics, such as latitude, climate, exposure, etc. Biomass, geothermal or hydro are even more area specific. This means that when we wish to utilize renewable energy we must look for the best available option at a local level. This could be biomass from the surrounding woodlands for one place, micro-hydropower and solar power for another, a small wind turbine for the third. It also means that small scale generation suited to the available but changing options will be more feasible. This would also allow scaling down the technologies and avoiding transportation over unsustainable distances, as is often the case with biomass.

However, while the technology associated with such systems is continuously improving, sustainability cannot be achieved by simply relying on efficiency improvement. A change in our way of thinking is needed. In our globalized world consumers have become detached from production: we consume food from the other side of the world, wear clothes that were manufactured on other continents and are unaware of the sources of our energy and what its use truly means for the planet. When consumers are so far, both literally and figuratively, from the point of production, how could we expect a more conscious attitude toward energy use?

Rural communities with knowledge about their locality are best able to invest in these small-scale energy systems. The goal of community energy is not just to achieve zero carbon energy generation, but to bridge the gap between production and consumption in order to create more informed behavior as the optimal path to achieving long term sustainability. Community energy projects may also help communities become energy independent and control prices by preventing or mitigating volatility caused by dependence on global oil and natural gas, which have seen sharp price rises and drops, thus furthering the community's resilience. Even if energy independence is not reached, for example, due to state regulations that don't allow direct use of the generated energy, these schemes could still save money for the communities on the long run and help them invest in development initiatives that could lead to higher living standards and a healthier environment. They are also an opportunity to create employment locally and engage with low carbon technologies and climate change impacts on a more easily understood scale. In less developed areas, small-scale community-owned energy systems could be the only way to provide basic services to residents, and by doing so, help in reducing poverty. In all cases, it helps to decrease the number of people living in fuel poverty. The term is used to indicate households whose energy costs are higher than the national average and would be left living below the official poverty line if they were to try to meet those costs, which in 2015 meant 11% of England's households (Department for Business, Energy and Industrial Strategy, 2017).

To summarize, small-scale community energy systems can help communities increase their independence, resilience, and awareness of environmental issues, while bolstering communication between residents and strengthening community ties. If we conserve and take care of the environment, it can provide us with healthy lives, income and livelihoods, cultural space and recreation. But no one can manage the environment by themselves. It falls to communities to take responsibility and care for their surrounding environment and by doing so to be an example to others.

3 Brief History of Community Energy in the UK

The community energy sector is relatively young in the UK. While various co-op enterprises have emerged in the first half of the nineteenth century, there were very few in the energy sector until the end of the twentieth century, when the first organizations started to appear (Harnmeijer, 2017).

The movement's origin could be traced back to the growing concerns regarding climate change and population growth that have led to the emergence of various environmental groups. Concerns were amplified during the oil crises of the 1970s and attention turned toward energy resources and possible ways to increase fuel security. Media coverage over the years about the various IPCC reports, commitments to the Kyoto Protocol and the publication of An Inconvenient Truth and the Stern Report have raised awareness on the issue of climate change and increased people's wish to tackle it (Cowtan, 2017). Despite this, renewable energy had not been a major focus of the government's agenda and was still "not perceived as an industrial opportunity" (Harnmeijer, 2017, p. 9). Nevertheless, in 1990 the British government launched a support mechanism called the Non-Fossil Fuel Obligation (NFFO) that increased the amount of electricity producers could receive if the electricity was produced from renewable sources (Cowtan, 2017). While at the beginning it was aimed at nuclear power, later it was extended to incorporate renewable energy as well. This could be regarded as the first government policy to support renewable energy. It helped form one community energy scheme: the Baywind Energy Co-operative (Cowtan, 2017). In 2001, the government introduced Renewable Obligation Certificates (ROCs) to help reach their target for generating capacity, which didn't happen under NFFO, by requiring electricity suppliers "to purchase electricity that was generated from renewable resources" (Cowtan, 2017, p.28).

At the turn of the twenty-first century, a Scottish development organization established an office for community energy in the large and scarcely populated country of Scotland. In 2007, this office was restructured as Community Energy Scotland (CES), becoming one of the most important focus points of the sector and helping to launch many projects. By the end of 2008, the total installed capacity owned by communities had reached 27 MW, out of which almost half could be attributed to Scottish projects (Harnmeijer, 2017).

However, the financial incentive created by ROC still proved to be inadequate for small-scale projects, as the initial costs required from communities to invest were still too high. This changed with the introduction of the feed-in-tariff (FIT) in 2010. Its goal was to offer cost-based compensation to electricity producers and ensure long-term contracts resulting in higher financial security for community energy organizations, while making projects financially viable. Thanks to this, the community energy sector has seen a sharp rise in schemes (Cowtan, 2017). However, in 2015, it was announced that the government would make significant changes to many of the supporting mechanisms, including substantially reducing the FIT (CEE, 2017). As a result of this decision, according to Harvey (2016), "key targets for renewable energy by 2020 were unlikely to be met." According to the government this was decided because the amount of money to be paid to renewables by 2020 was projected to exceed the previously agreed limit, leading to bill payers having to pay more (Gov.uk, 2015). The situation hasn't changed since then, even though polls continue to show high support for renewable energy projects (Harvey, 2016).

4 Case Studies of Community Energy Projects

The four projects chosen as case studies are located in North West England, in Cumbria county (Fig. 1).

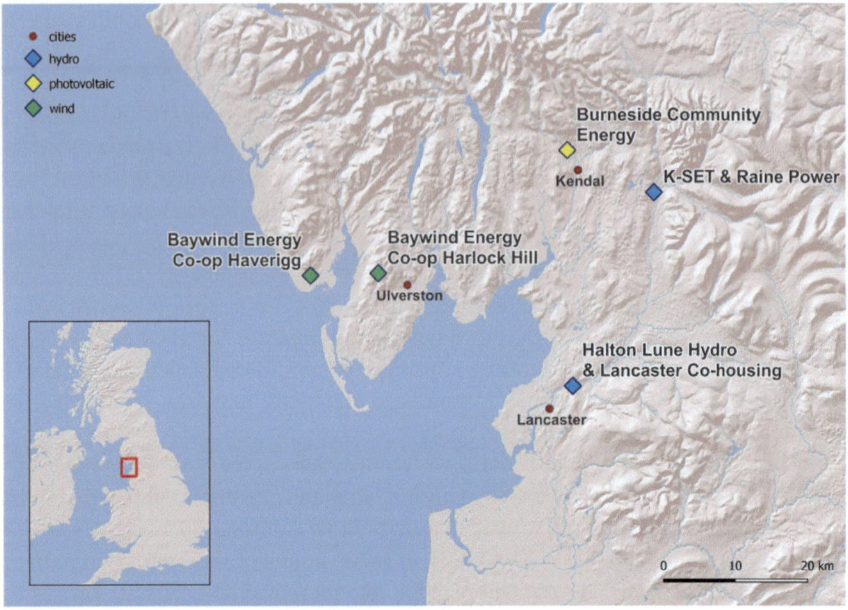

Fig. 1 Map of four case study sites and location of Cumbria on the national map

4.1 *Baywind Energy Co-Operative and Energy4All*

Baywind was founded in 1997, making it one of the oldest energy co-operatives in the UK and the first in the country to own wind turbines in the country. They have followed successful community energy models, particularly Denmark's, as it was a Danish company that first recognized the opportunity for wind energy created by Britain's gently hilly countryside, particularly in Harlock Hill, in Ulverston, which later became Baywind's main site and where five out of its six turbines have been installed.

The share offer happened in three stages. The first raised £600.000, which was enough for the first turbine. It was an act of faith on the part of shareholders, because despite having the permissions and agreement in place, at the beginning there was not yet anything in the ground. Transporting the blades and the pillars to the site was also challenging: walls along the road had to be demolished, and then repaired.

After about 20 years, the turbines got decommissioned in 2016 because there had been a change in government policy regarding co-ops and they couldn't renew their permit. Therefore, after years of planning, the site was taken over by High Winds, their successor and a BenCom, which installed two 2.3 MW turbines. Baywind currently owns a share in a turbine at a different location in Haverigg, but has no turbine on its own. Decommissioning turbines is expensive, but Baywind fortunately was able to sell them to an Irish scrap dealer who most likelyresold them to be installed elsewhere or to be used as spares.

At one point, the owner was keen on selling his land, but Baywind did not take advantage of that. In retrospect, that was most likely a mistake as owning land could give more opportunities for a community organization, says Baywind's director.

In 2017, Baywind, High Winds and Energy Prospect co-operatives, together with an ethical investment company named Thrive Renewables, acquired an additional three turbine wind farm on Harlock Hill, resulting in a current total of five turbines with the total installed capacity of 11.5 MW on sites owned by community organizations.

Baywind also has its community fund, which has donated some funding to local homes and public facilities to put up PV panels, insulate or install LED bulbs. Due to being a coop and not a BenCom, and having to overcome some expensive technical problems, the amount they were able to provide the fund was relatively small. With High Winds this should be different and the financial benefits for the community should be substantial.

Wind turbines in the countryside are a very logical choice: turbines require small areas for their operation and the space around them can be used for agriculture or livestock grazing. The animals walk around freely and are not bothered by the facilities. Farmers looking for extra income might be able to lease their lands, while still keeping their original activity going. This is the case with Harlock Hill as well, where the land is owned by a local farmer who still grazes sheep. However, despite the many benefits, the small area needed and the possibilities for the coexistence of agriculture and renewable technologies, onshore wind turbines have become an unwelcome sight in the UK as many started to view them as something that ruins the scenery. The backlash against onshore turbines led to a ban on future projects, even though the UK has among the highest potential among the European nations for wind power generation (EEA, 2009).

Baywind's 1,217 shareholders are now discussing how to invest their unused capital. They wish to avoid acting simply as an investment company, but how to proceed from now on is yet to be decided.

The experience of Baywind suggests that community energy projects offer a continuous learning process. Unexpected opportunities may arise, such as available land, or a scrap dealer who takes away decommissioned turbines. Or problems may need to be solved, like technical issues, or the change in government policy that makes repowering impossible. Baywind members realized that there are other community groups who might like to do something similar, but lack guidance and expertise. To address the absence of similar initiatives Baywind members established Energy4All in 2002, an umbrella co-operative linking 23 independent renewable energy coops and BenComs (Energy4All).

Each project has a board of directors, but few can afford to hire employees. Energy4All can offer support, helping communities develop their own energy projects, raising funds for them, assisting with daily administration and serving as the point of contact for shareholders (Energy4All annual report, 2015).

Energy4All manages projects not just in rural environments, but in large cities as well: Edinburgh Community Solar Co-operative has installed PV panels on 24 roofs across the city, with a total installed capacity of 1.38 MW. The panels are located on top of community buildings, such as schools, leisure centers and community centers (ECSC). Marks and Spencer has created a BenCom under the name M&S Energy Society with the aim of installing PV panels on the roofs of nine of its stores (M&S Energy Society). They are also administered by Energy4All. With these projects E4A has successfully ventured into new waters, opening up the possibility to similar projects in the future (Energy4All annual report 2017).

4.2 Halton Lune Hydro and the Lancaster Cohousing

The River Lune has powered mills along its river bank as far back as the thirteenth century. Today only one mill remains, the Halton Mill, although not in its original capacity anymore. Nevertheless, the river has many areas that could potentially house a hydropower project. A community association in Halton decided they would like to put a micro hydropower dam on the river, close to the mill, so they created the Halton Lune Hydro (HLH) BenCom to benefit the villagers as well. Some 450 people, mainly local residents but some from throughout the UK, have invested in the project. The hydro scheme has two turbines, each generating 100 kW; the first started to operate in 2014, the second a year later, although this one generates only 60 kW currently.

While the HLH initiative was coalescing another group of people visited the area looking for a suitable location for a cohousing complex. The idea behind cohousing is that while everyone has their own living space and privacy, certain facilities are shared among the residents, hence reducing the need for large houses and individual equipment and facilities. The two groups met each other during a site visit and decided to collaborate, investing jointly in a private wire or micro-grid. The generated electricity from the turbines feed into this private wire and supplies the Lancaster Cohousing with energy: the 160 kW can supply about 300 homes, but they only have 41, so the rest enters the national grid. During the development another BenCom, called MORE (Morecambe Bay Community Renewables), joined the initiative, offering to install solar panels (PV) on the rooftop of the cohousing buildings. While it would have been possible for the residents of the cohousing to do this on their own, they wanted to work with other community groups as well, so they now house 90 kW of PV panels.

The lifestyle of a cohousing encourages sharing and connection with the neighbors through common facilities, which are in this case a large dining room, where residents are encouraged to cook for everyone once a month and share meals with others as often as possible, a children's playroom for the residents' children, a laundry room to decrease the need for washing machines in every home and a workshop containing tools and equipment donated by all the residents, who can borrow freely. The cohousing also has its own shop in order to reduce the need to drive to town. The store sells organic and mostly local products. The building intentionally has limited parking spaces to discourage car use. Instead they have a car club with six vehicles, of which two are electric. The cohousing has 35 units that are passive, energy efficient, houses and form the cohousing complex, and six more that are built separately by request of future residents, but are passive houses as well.

When they bought the property the cohousing had to keep the Halton Mill as well, which by that time was renovated. It was turned into an office space with various facilities, including a café, a meeting place, an arts and crafts studio, seminar rooms, yoga classroom and a shared workspace. The cohousing together with the mill have their own unique energy system. The most important are the turbines a little up on the road, as well as the PV panels on the cohousing's roof. On the mill's roof there are solar thermal panels that preheat the water before it enters into the biomass (woodchip)-fueled heating system. This means that the cohousing and the mill have their own small district heating network. The energy the cohousing and the mill use are from 100% renewable resources. The coexistence of these two projects suggests that such opportunities might rise elsewhere too where groups sharing similar objectives might join together to reach their goal; it also shows that it is possible for small communities to become energy independent and supply their energy needs from completely renewable sources. There is one more advantage this hydro project has: in many cases a project's continuity is in jeopardy once the management retires from the project. In this case, rather than having a main contractor, the management of the hydro project had decided to work with various stakeholders, resulting in having no one that fully understands the project outside of the management. However, thanks to the cohousing there is a reservoir of expertise around and younger residents are already getting involved in hydro management. With a complex technology such as a hydroelectric facility many things can go wrong; having participants who understand both the technology and this particular site will help with the long-term survival of the project.

The HLH also aims to give back to the community and not just repay the investment to the shareholders. Surplus funds go into the Halton Lune Trust, which then redistributes the money between projects that aim at improving sporting and other recreational activities, and those that try to improve energy efficiency of local houses and promote carbon reducing technologies (Halton Lune Hydro). The amount of funds the charity receives each year varies depending on hydro conditions: according to regulations they are only allowed to siphon off a designated amount of water from the river, which negatively impacts the performance of the hydro. In order to protect the fish and ensure their safe passage they were required to erect a net; however that limits the water intake, as the mesh is only 10 mm. This results in the output of 130 kW, instead the intended 160 kW. To further protect the fish a fish passage was built and cameras installed for monitoring. They are also required to trap 5–10 percent of the fish population and weigh and measure them as well as check for diseases. They hope to build trust with the Environment Agency to allow them to increase the size of the mesh.

There are also unexpected events, such as floods that damage the turbines or the infrastructure, and droughts that reduce the energy generation of the turbines. Because of the changing climate, the number and severity of such events have increased. Storm Desmond hit the UK in 2015–2016, which resulted in severe floods in multiple locations across Northern England. The hydro was also impacted, as the water level of the river broke records and the turbines were inundated. This caused the project to stand down for two months until the building and the turbines were repaired. In April and May of 2017 drought caused the turbines to stop generating as the water levels were too low. The loss of generation also cuts back on revenues. To repair the damage to the powerhouse or the turbines themselves costs a lot of money, and after the repairs revenues go first to shareholders, and only then to the community fund. Such unforeseen events will impact the amount of money the shareholders receive as well. This means that these energy projects can be financially risky investments. While they tend to have a much higher return rate than do banks, there is always the possibility of not receiving the promised amount in a given year. Of course, the shareholders are aware of this and the fact they still gladly invest implies that there are other benefits to them: these projects have high positive impacts on the environment and the community.

4.3 K-SET and Raine Power

Another micro hydro project on River Lune, the same river on which Halton Lune Hydro is located, started with a small village initiative called the Killington and District Sustainable Energy Trust (K-SET). The village residents were keen on addressing climate change in their own way. As a first step they did a carbon footprint analysis, by which they learned that the amount of carbon-based energy they were using in a year is three times that of the national average. As a small farming community of about 160 residents this of course makes sense: farmers need tractors and other electric equipment to do their job and they also need to drive cars to get to work or school. Nevertheless, residents wanted to do something about this so they created K-SET in 2009. Their first project was not energy generation but increasing energy efficiency by insulating houses to decrease dependency on carbon. Later many invested in air source or ground source heat pumps.

The next step was to invest in an energy generation technology on the river. They decided to install a 35 kW Archimedean screw that would allow fish to safely pass through it. K-SET had raised the money for the planning and got the permissions, but they soon needed more funding for construction, so they turned to Energy4All. With their help a new BenCom, Western Community Energy Society (better known as Raine Power) was established in 2015 (Raine Power). From that point on Raine Power handled fund raising and construction for the project.

The journey to a working hydro had been very challenging and the community organization's travails are still ongoing. It is an old initiative, yet after all these years, the constructions are still incomplete. Concerns from residents about possible noise from the screw had to be addressed by putting a lid over it, even though any possible noise the screw could generate will be hidden by the noise of the weir. However, this meant further permissions for which they had to wait and stop construction, not to mention adverse impacts on finances. The construction halt caused the greatest concerns: because of the movement of the trout they were not allowed to work on the river after 30 September. However, because the work couldn't start until July they were unable to finish by the deadline. And because of further regulations, they couldn't leave the construction site as it is, it had to be restored to its original condition. This meant that the work done between July and September was in vain and the next year they had to start from the beginning. This, as well, had a great impact on the finances. Harsh weather and the high water level destroyed the path to the power house in November 2017, further delaying completion. The power house and the turbines inside were left unharmed as they are well insulated to protect nearby residents from the noise of the turbine. Gaining the right of way to lay down the cable was also very challenging and has slowed down the completion significantly. All these have caused costs to soar, leading to fears that the project might not be able to provide any funds to the community after they pay back their shareholders, and that they may have to wait longer to receive interest. Furthermore, if they are unable to finish by February 2018, they will lose the feed-in-tariff. The project is nearly complete, but before laying down a cable, the final step, a storm hit, which washed away part of the construction. They expressed hope that they could restore it and finish in time. After all these hardships, they still haven't given up and hope to reap the fruit of their persistence.

4.4 Burneside Community Energy

This community project is the newest of the four, but that's not all that sets it apart. In 2009, a group formed to look at the possibilities regarding renewable energy generation in Cumbria. They had sent out a survey to 300–400 households to determine how people in the extended community of Cumbria felt about renewable energy. The 296 replies that came back showed very positive attitudes: people were interested and they wished to invest. Seeing this, the group started to examine the possibilities of a community energy scheme. Talks with a consultation company led them to locate a perfect site for a hydro project at Killington reservoir, but this scheme was eventually implemented by another community organization, Community Energy Cumbria (CEC), with the help of Cumbria Action for Sustainability (CAfS), a Penrith-based charity organization that tackles climate change by promoting various energy saving methods in buildings, low carbon lifestyles and reduced use of fossil fuels. They achieve this by providing information, advice, and training, and by conducting site visits and educational events. Their activities center on such areas as draught proofing, preparing and increasing resilience against floods, energy saving at home, and community energy projects. The Killington reservoir project, which was officially opened in January 2017, provided both CEC and CAfS with increased knowledge of how to launch and develop community energy projects and bolstered trust toward them. Seeing their success, the chairman of the James Cropper PLC paper mill located in Burneside received the permission of his company's board and offered their roof for installing PV panels free of charge, testifying of very positive corporate social responsibility towards the local community.

However, time was pressing because with the end of 2015 the feed-in-tariff (FIT) was to be significantly reduced and tax relief was to be abolished. In October 2015 they created Burneside Community Energy (BCE). The organization was officially registered as a BenCom one hour before the share offer was to open. In just four days, the money for the project was pledged. However on 5 December Storm Desmond hit the town, causing severe damage. Nevertheless, by the 16th the project recorded 250 kW of installed capacity from over 100 panels, a notable achievement.

As one of the founding directors of BCE noted: "projects like this don't have risk capital, they have risk individuals." The chairman of the paper mill had put down the deposit for the PV before opening the share offer, taking a great personal risk. In the beginning, it would have been safer to recognize that the project was not likely to be achievable in just two months (from registration to installation), since if it had failed the news would have traveled far.

BCE has always viewed this as just the beginning of the journey. The James Cropper mill has a wastewater storage tank where they store the byproducts from waste paper recycling and in theory it was suitable for use in a local district heating system. However, the houses in Burneside are not suitable for such a system as they are not properly insulated and the pipes are too old. The water would not be warm enough to supply heat if it was sent through leaky pipes into badly insulated buildings. Therefore BCE decided to focus on a new housing development project that is still in the planning stages. There are already two sites designated for housing. If these homes are built to be highly insulated, with high levels of airtightness and low heat demand, they would be appropriate for using the heat from the wastewater. Even so, however, 2–3 kW PVs on the roofs and heat pumps in the wastewater tank and the river would be necessary to make the system work, making it very complex and expensive, although not impossible. Innovative technologies could be introduced as well, such as battery storage and electric vehicle charging. As the complexity of such a system is beyond the capacity of most community energy groups, such an initiative would require the creation of an energy supply company. The James Cropper paper mill would also need to upgrade its capacities and transform itself as an energy provider as well as a mill. If such an ambitious plan is realized, Burneside would have a very unique energy system for the UK.

5 Lessons Learned

Community energy is a topic that seems fairly simple but interviews and observation reveal it to be quite complex. It can encompass energy generation and distribution, but energy efficiency and educational activities are equally important components.

Community projects give priority to people living in the community, but in certain cases they might open the share offer to others as well. Depending on the minimum amount for investment local residents may find participation too expensive or may simply be uninterested. The small number of people living in a community may also be limiting factor, requiring that groups open the share offer to a wider audience.

This could lead to the involvement of shareholders living far from the site, leading some to argue that a community project where the shareholders don't belong to the community should not be considered a community project. Yet it could also be argued that one could belong to a community not based on geographical location, but on shared principles. Potential shareholders might not have the opportunity to invest in a local project, but nevertheless feel highly motivated to support low carbon energy production.

Investing in community energy schemes can involve high financial risk. During a share offer it is stressed that one should only invest an amount that could be lost without having dire effects. Investments are a long-term commitment; investors should understand that their money cannot be taken out at any time. The ability and timing of withdrawal depend on the financial state of the scheme and potential tax returns. For example, in the case of Halton Lune Hydro, people only received tax relief if they left their money in the project for at least three years. Furthermore, while returns to investment are estimated, unexpected events may reduce the amount or even delay them. If the project does well, the possible income could be very high, but its opposite could happen just as well. Nevertheless, many people seem to be interested in these schemes and if they have the money, they will invest. This implies that people join for other reasons beside financial benefits: environmental and community consciousness play a greater role.

Setting up projects is very hard. There are numerous possible problems that can result in setbacks, as was shown by Raine Power's hydro project. Government policy and incentives, such as subsidies, FIT and tax reliefs are very important factors for incentivizing community energy projects. Without them, the financial burden on shareholders would most likely make these projects untenable. Unfortunately, the government pared back its support for such projects in late 2015. Among many reductions, the continuous cuts on the FIT were most impactive for the sector. This jeopardizes the future of community energy projects, yet the sector shows resilience through innovation and collaboration with other communities (CEE, 2017). According to Baywind's director, 20–25 years ago community energy projects were rare therefore receiving permission was easy and applying for a 20 year long operating time permit seemed more than enough. If they had to renew their license, they could just reapply. However, government policy toward community energy has changed in recent years, and in the case of onshore wind turbines they have been banned. Projects that are to be over soon most likely will not receive permits barring a sudden change in government policy. If this wasn't problematic enough, the future of each project depends on the enthusiasm of the people involved: whether they wish to renew it, end it, or further develop it. Community energy projects can be regarded as a balancing act, with organizations relying on trust, communication and diplomacy in order to keep everyone on board and the project moving.

One way to cut back costs and strengthen the involvement of the community is to seek out shareholders with expertise. If people are involved in the development, management or maintenance of a project they are more likely to feel it is theirs and do more to keep it going.

Once a scheme is up it will continue to run until the technology and/or permissions expire. Nevertheless, projects will most likely encounter snags along the way that will need to be solved. Technical problems such as disabled cables due to malfunction could happen anytime, as happened to Baywind, but environmental impacts can cause harm as well. Climate change affects the weather and in return that impacts the technology. The rivers flood and water damages the hydro plant, or there are droughts and the turbines cannot generate as much as necessary, as was the case for Halton Lune Hydro. The incidence of extreme weather events is rising and that has negative impact on the projects.

The existence of community energy projects is thanks to a handful of very dedicated individuals who work tirelessly to make them happen, most of the time without financial compensation. This also implies that unless they are able to pass on the management duties to equally dedicated people, the future of these projects may become uncertain.

6 Conclusion

Community energy is one way to tackle climate change by decentralizing power generation and by utilizing renewable energies. However, it can also be an important tool in increasing the resilience of communities against fuel poverty and lowering electricity prices. It can help strengthen community ties and lead to an increase in income for the area by supporting local jobs and other development projects.

Energy schemes may face multiple problems during their development, but even after that as well. Technical problems will always arise and with the changing climate the impact of extreme weather events will endanger operations. Community energy in the United Kingdom is young and as such, it is still fragile (Cowtan, 2017). The recent cessations and reductions of government support further weakens it. However, community projects are bottom up movements and as such, they depend more on individual leadership and dedication than on regulations from above. The help in the form of positive government policy is very much needed, but until that time comes, community energy looks likely to continue.

The four case studies introduced in this chapter are projects in England. It is important to note that the Scottish government's support toward renewables and community energy projects has been much more favorable at the time of writing this chapter despite the changes in government policy in England (Cowtan, 2017 and CEE, 2017). The Welsh government is also lobbying to change government policy. And even in England, co-ops are trying to come up with ways to increase the feasibility of projects by working together with other groups or businesses. Community energy will continue.

Acknowledgements The author would like to express her utmost gratitude to the members and directors of each project introduced here for their kindness and support in developing this study.

References

Alvial-Palavicino, C., Garrido-Echeverría, -, Jiménez-Estévez, N., Reyes, G., Palma-Behnke, L., & R. (2011). A methodology for community engagement in the introduction of renewable based smart microgrid. *Energy for Sustainable Development, 15*, 314–323.

Community Energy England (CEE) (2017). *State of the sector report: A study of community energy in England, Wales and Northern Ireland*. Retrieved from: https://communityenergyengland.org/pages/state-of-the-sector-report. Accessed: 2018.01.20.

Cowtan, G. (2017). *Community Energy: A guide to community-based renewable energy projects*. Green Books.

Cumbria Action for Sustainability: http://www.cafs.org.uk/. Accessed: 2018.01.20.

Community Energy Cumbria: http://www.communityenergycumbria.co.uk/. Accessed: 2018.01.20.

Department for Business, Energy and Industrial Strategy (2017). *Annual fuel poverty statistics report 2017 (2015 Data)*. Retrieved from: https://www.gov.uk/government/collections/fuel-poverty-statistics. Accessed: 2018.01.20.

European Environmental Agency (EEA) Technical Report (2009): *Europe's onshore and offshore wind energy potential —An assessment of environmental and economic constraints*

Energy Networks Association. (n.d.) *The rough guide to engaging communities in network innovation*. Retrieved from: http://www.energynetworks.org/assets/files/news/publications/ENA%20Engaging%20Community%20in%20Network%20Innovation.pdf. Accessed: 2018.01.20.

Edinburgh Community Solar Co-operative (ECSC): http://www.edinburghsolar.coop/. Accessed: 2018.01.20.

Energy4All. (2015). http://energy4all.co.uk/. *Energy4All annual report 2015, Energy4All annual report 2017*. Accessed: 2018.01.20.

GOV.UK (2013) Community Energy. Retrieved from: https://www.gov.uk/guidance/community-energy. Accessed: 2018.01.07.

GOV.UK (2015) *Changes to renewables subsidies*. Retrieved from: https://www.gov.uk/government/news/changes-to-renewables-subsidies. Accessed: 2018.01.22.

Harnmeijer, Anna (2017) *A short history of community renewable energy in the United Kingdom*. In press for: Holstenkamp, L. & Radtke, J. (Eds) Handbook on Energy Transition and Participation, Springer Verlag, Wiesbaden.

Harvey, Fiona (2016). *Just 10 new community energy schemes registered after Tories cut subsidies*. The Guardian. Retrieved from: https://www.theguardian.com/environment/2016/sep/12/just-10-new-community-energy-schemes-registered-after-tories-cut-subsidies. Accessed: 2018.01.20.

Halton Lune Hydro: http://haltonlunehydro.org/. Accessed: 2018.01.20.

K-SET: http://www.k-set.org/. Accessed: 2018.01.20.

Lancaster Co-housing: http://www.lancastercohousing.org.uk. Accessed: 2018.01.20.

M&S Energy Society: https://www.mandsenergysociety.com/. Accessed: 2018.01.20.

Raine Power: https://rainepower.coop/. Accessed: 2018.01.20.

Issues and Challenges in Implementing Social Forestry Programs in Customary Communities: A Case Study of *Hutan Adat* in Indonesia

Andarini Sertianti and Jane Singer

Abstract Social forestry programs have become an increasingly popular approach to mitigate deforestation and biodiversity loss in forest-rich countries. These programs emphasize the involvement of local communities or indigenous peoples in establishing sustainable forest management systems in their respective regions. The aim is primarily to prevent further deforestation, while promoting rural development, environmental protection and the welfare of forest-dependent communities. The Indonesian government has been implementing several social forestry schemes to counter the adverse impacts of rapid agricultural expansion and infrastructure development on the well-being of local communities. This chapter examines one of these, the customary forest scheme (*hutan adat*), which recognizes customary indigenous land rights and hands over forest management to indigenous people. Past research has shown that recognition of indigenous land rights can ensure the revitalization of traditional forest management which, in turn, may help safeguard the sustainability of forests and community. However, some doubts have arisen as to whether indigenous communities are ready and capable of securing the long-term sustainability of the program. This chapter will discuss cogent issues surrounding the implementation of customary forest schemes in Indonesia by examining the experience of a recognized customary forest in Rantau Kermas village in Jambi province in central Sumatra.

Keywords Social forestry · Customary forest · Indigenous land rights · Forest governance · Community-based forest management · Indonesia

A. Sertianti (✉)
Independent Researcher, Kyoto, Japan
e-mail: andarini.s@gmail.com

J. Singer
Department of Global Studies, Kyoto University of Foreign Studies, Kyoto, Japan
e-mail: j_singer@kufs.ac.jp

M. Yoshizumi et al. (eds.), *Navigating Local Sustainability in Food, Community, and Innovation*, https://doi.org/10.1007/978-981-96-8963-7_6

1 Introduction

Over the past few decades rapid rates of deforestation and forest degradation, especially in developing countries, has drawn major attention worldwide. Particular concern is being voiced about the rapid loss of forest cover in tropical countries. Tropical rainforests, which are regarded as the richest and most valuable types of ecosystem on the earth's land surface, account for 41 percent of the total tropical forest cover worldwide (Angelsen & Kaimowitz, 1999). Taking into account the importance of tropical rainforests, losing these forests has been predicted to result in various adverse consequences such as an increase in carbon emissions as well as a loss of biodiversity and sources of livelihood. Angelsen and Kaimowitz (1999) noted that more roads, higher agricultural prices, lower wages, and a shortage of off-farm employment generally lead to more deforestation. Among these factors, agricultural expansion has been found to be the main driver behind tropical deforestation cases around the world (Geist & Lambin, 2002). In Africa and Asia, both subsistence and commercial agriculture contribute almost equally to land change in forest areas, although it is predicted that commercial actors might play an increasingly larger role in agricultural expansion in the forest (Hosonuma et al., 2012).

Deforestation prompted by massive agricultural expansion has also become a phenomenon in Indonesia. Forest cover data published by FAO indicates that the forest area in the country has drastically declined from 118.5 Mha in 1990 to 91.0 Mha in 2015 (Food & Agriculture Organization of the United Nations, 2015). Although varying factors may account for forest change, the imbalance between global demand and the production of Indonesian timber and oil palm, which has led to illegal or non-sustainable agricultural expansion, was claimed to be the leading cause of the sharp decline of forest area from the mid-1990s to 2015 (Tsujino et al., 2016). Forest Watch Indonesia suggested that the rate of forest loss in Indonesia between 2009 and 2013 alone was approximately 1.33 Mha per year (FWI, 2014). If this rate continues, natural forest cover in several provinces in Indonesia, including those in Sumatra such as Riau and Jambi, may be completely lost by 2033 (FWI, 2014).

Severe deforestation and forest degradation does not only bring adverse consequences to environmental health but also to the forest-dependent people who live in or adjacent to forest areas. To these people, forests provide many services, from provision of forest products for their livelihoods to environmental services. Deforestation and forest degradation can directly and indirectly impact the economic activity of forest-dependent people and put their livelihoods and cultural integrity at risk (Angelsen & Kaimowitz, 1999). As many as 1.6 billion people around the world depend on forests for their livelihoods to different degrees (United Nations Environment Programme & GRID–Arendal, 2009). It has been estimated that the number of Indonesian people who depend on forests for their livelihoods may be as many as 65 million (FWI/GFW, 2002). In addition, approximately 25,800 villages, or more than 36 percent of all villages in Indonesia, are located inside or adjacent to forest areas, and 98 percent of these villages rely on agriculture for their livelihoods (Badan

Planologi Kehutanan, 2007 as cited in Nugroho et al., 2017, Direktorat Jenderal Planologi Kehutanan, 2009 as cited in Nugroho et al., 2017).

Many forest-dependent people live a traditional lifestyle that includes shifting cultivation of rice and other food crops while harvesting and selling timber and non-timber forest products such as rattan and honey (FWI/GFW, 2002). Many of them live in poverty and their well-being is often virtually disregarded by policymakers to favor land conversion for agricultural purposes, which is perceived as more lucrative for regional economic development. Additionally, forest conservation approaches applied in the country are mostly aimed at creating unperturbed forest areas free from human activities. While it seems ideal, this idea overlooks the fact that many people have been relying on forests for a long time. Consequently, this notion has created a conservation policy that harms the interests of forest-dependent people. As forest-dependent people try to protect their lands, disputes and conflicts over land often occur between them and local government or companies who wish to claim the forest areas for other uses. These conflicts, more often than not, favor those with strong political and economic power and result in deprivation of access to forest areas for local people. Being deprived of access to forest has consequently impoverished many forest-dependent communities, who generally have little to no political power and information regarding tenure security, and it has made them among the most vulnerable groups in Indonesia. The conflicts and massive land conversion has also exacerbated the environmental problems induced by deforestation and forest degradation. Additionally, poor economic conditions and lack of awareness about environment have, in some instances, prompted forest-dependent communities to cut timber in an unregulated and unsustainable way for quick cash or to clear their forest to open land to grow more profitable crops. Such actions may produce short-term economic benefits, but they will harm the environment and their well-being in the long run.

In an attempt to mitigate the problems caused by deforestation and alleviate the impacts of land conflicts, while ensuring the well-being of forest-dependent people, the Indonesian government has introduced several schemes under its Social Forestry (*Perhutanan Sosial*) program which generally aim to share the responsibility of forest management by government with the people. Rules and regulations regarding the Social Forestry program are included in national law P.83/MENLHK/SETJEN/KUM.1/10/2016. Based on the law, there are six categories of the Social Forestry program in Indonesia. These include: community forest (*hutan kemasyarakatan*), village forest (*hutan desa*), community plantation forest (*hutan tanaman rakyat*), collaborative forest management (*kemitraan kehutanan*), smallholder private forest (*hutan rakyat*), and customary forest (*hutan adat*). Each category targets different types of communities as well as forest characteristics. Each also has different regulations and benefits for the communities that are involved. Indonesia is not the only country where social forestry is undertaken as one of the approaches to mitigate deforestation and its impacts to local people. In fact, Social Forestry, also referred to as Community Forestry or Community-based Forestry, is increasingly becoming a popular option to establish a sustainable forestry system in developing countries where centralized forestry management by the national government has often failed

(United Nations Environment Programme & GRID–Arendal, 2009). Despite the promise of these schemes, however, some have questioned whether social forestry programs can be implemented successfully and sustainably in Indonesia. Besides, as the program encourages a rather high degree of involvement from local people, the question of local communities' readiness and preparedness inevitably emerges.

This article will discuss the issues and challenges that arise following the implementation of social forestry programs in Indonesia, specifically in customary (*adat*) communities, under the customary forest (*hutan adat*) scheme. Customary forest was chosen as it has more distinctive features and regulations compared to other social forestry schemes in Indonesia. Customary (*adat*) communities in Indonesia generally consist of peoples who are indigenous to their land, and thus have a long history and traditions in regards to owning and managing the forest areas within their land. In addition, as customary communities often reside in rural areas with highly limited access to infrastructure and information, a large number of these communities do not have legal deed to property nor their land, despite claiming long-time occupation. Amidst the long-term land conflicts in many parts of Indonesia, the handover of legal rights over customary forest from government to customary communities seems like a welcome step. However, as more customary communities receive legal recognition, the aforementioned questions regarding their preparedness and the program's sustainability also follows. This chapter will introduce a case study of the implementation of social forestry in Rantau Kermas Village in Jambi province in central Sumatra. The aim is to shed light on the issues and challenges surrounding the implementation of the program that have occurred and may arise in the future. The methods used in this study include a review of relevant literature as well as interviews and focus-group discussions (FGD) conducted with the residents of Rantau Kermas Village by the first author in November 2017.

2 Social Forestry as a Concept

The term social forestry, often used interchangeably with community forestry or community-based forestry, first came to prominence in the mid-1970s (see Table 1) (Arnold, 2001). The United Nations Food and Agricultural Association (FAO) in 1978 defined community forestry as "any situation which intimately involves local people in a forestry activity" (Arnold, 2001). As the definition suggests, community forestry covers a very broad range of activities and linkages among people, forests, and forest products, but it has largely been introduced in forest-rich developing countries (Arnold, 2001). Today, many forms of community forestry are recognized in many different countries, following the adaptation of forestry management to suit the local social, political, cultural and bureaucratic context (Gilmour, 2016). Similarly, different names such as social forestry and community-based forestry are also applied. Despite differences in terminology, it is largely believed that the names were originally adopted from community forestry and that they fundamentally share the same idea of community forestry. In this report, the term social forestry will be used, as it is a literal translation of the Indonesian term used by the Ministry of Environment and Forestry of Indonesia, *Perhutanan Sosial.*

The core concept of social forestry activities is that it is not a separate form of forestry but instead, a part of the process of adapting forestry and forest management to make it more responsive and relevant to the needs and interests of rural people in the context of broader societal and economic change (Arnold, 2001). The Ministry of Environment and Forestry of Indonesia (MoEF) defines social forestry (*perhutanan sosial*) similarly, as "a sustainable forest management system implemented within state forest areas or within customary forests implemented by local communities or indigenous and tribal peoples as key actors to improve their welfare, environmental balance, and socio-cultural dynamics in the form of village forest, community forestry, community plantation forest, smallholder private forest, customary forest and collaborative forest management" (2016) (see Fig. 1).

The main reason behind the emergence of the social forestry concept in many developing countries is the lack of involvement of local people in forest governance and forestry. When forest resources were abundant, unregulated 'open source' use was likely to exist (Arnold, 2001). However, as populations and economies grew,

Table 1 Key global forest-related development that have influenced the evolution of community forestry

Period	Events	Response
1970s	Fuelwood crisis	Initiation of forestry for local community development
	Failure of forest industry development model to sustain forests and meet community needs	Establishment of fuelwood plantations (generally top-down); many failed
1980s	Large-scale deforestation; environmental degradation	Pilot projects test CBF modalities in different settings to address environmental concerns
	Forest sector reforms: decentralization and devolution policies	Emergence of "people's participation" and bottom-up development
1990s	Sustainable development paradigm	Focus on SFM and livelihoods as CBF objectives
	Recognition of indigenous people's rights	Establishment of CBF regimes that formalize indigenous people's rights to manage forests
		Expansion of CBF across all regions
2000s	Globalization, trade liberalization	Growing interest in commercialization of wood and non-wood goods and services produced under CBF
2010s	Global policy focus on climate change, illegal timber and payment for environmental services	Additions to CBF objectives to address global policy interests

Source Gilmour (2016)

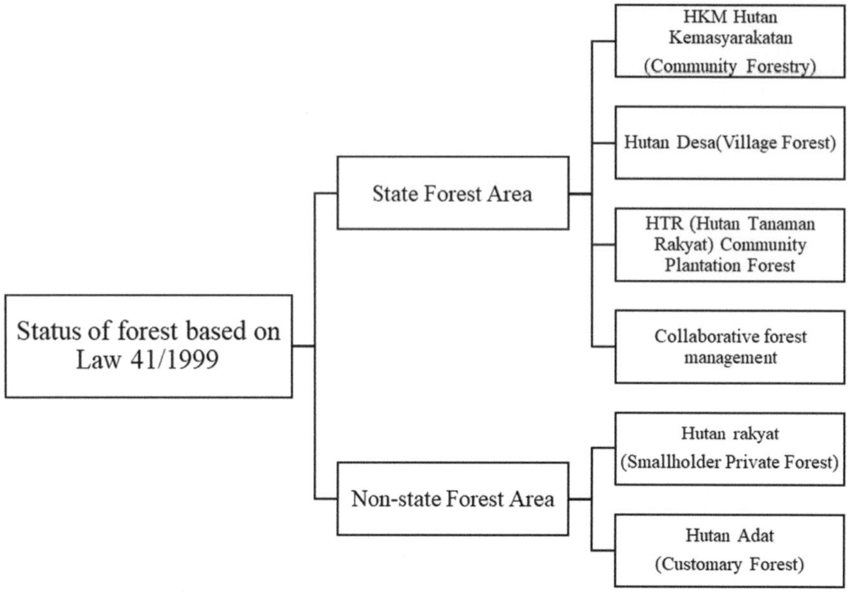

Fig. 1 Status of forest and social forestry schemes in Indonesia

pressures on availability of lands and resources have increased. Increasing use of forests has also led to progressive degradation. As a consequence, some form of control over rights to access and use has been imposed on local people.

Decision-making in forestry management in developing counties typically falls under the authority of the local or central government. However, sometimes the regulations formulated by the central government do not serve the interest of local people and the voices of those who live adjacent to or in forest areas are neglected. This situation makes the local forest-dependent people vulnerable to the often adverse impacts of changes in policy and regulations. In many of these countries, forest management led by central government results in the failure to promote sustainable management (United Nations Environment Programme & GRID–Arendal, 2009). When rapid resource-led economic growth accompanied by infrastructure development becomes the main priority of the national development agenda, and environmental health and the well-being of forest-dependent people are being compromised to make room for profitable economic activities such as large-scale plantations, it is not surprising that centralized forest management in many developing countries fails to ensure sustainability or human security for local people. In 2007, for example, in just one month, more than 500 incidents of conflict were reported between local communities and private interests attempting to establish oil palm plantations in Indonesia (Friends of the Earth et al., 2008 in United Nations Environment Programme & GRID–Arendal, 2009). In light of these concerns social forestry is often regarded as a measure for achieving positive environmental and community outcomes through the transfer of limited rights, discretionary powers, and capacity to local communities (Anderson et al., 2015). Thus it is often associated with decentralization and devolution.

The proportion of forest owned and managed by local communities in highly forested developing countries is estimated to be about one fourth of the total (Molnar et al., 2004; Scherr, White, & Kaimowitz, 2004; White & Martin, 2002). The research also indicates that this Fig. has doubled in 15 years since 1985 to 2000, and it is estimated to significantly increase in the next decades. Yet what actually makes social forestry such a popular and seemingly promising approach for sustainable forest management? Gilmour (2016) argued that the popularity of community-based forestry relies primarily on a narrative popularized by Ostrom in 1990 that local communities, when granted sufficient property rights over local forest commons, are able to organize autonomously and develop local institutions to regulate the use of natural resources and manage them in sustainable way. Additionally, in general, social forestry is highly regarded as an approach that combines forest conservation with rural development, community empowerment, and poverty reduction objectives (Anderson et al., 2015). In countries where the problem of deforestation and forest degradation is extensive and the living condition of forest-dependent people tends to be impoverished, social forestry may seem to be a win–win solution.

The concept typically involves local people as well as officials in forestry management, with the assumption that as local people have lived in the area for a long time, they possess knowledge about the specific local conditions of the forest such as which crops suit the ecosystem and a traditional resource management system that is able to sustain the community. The central government and outside experts may be seen as lacking this expertise so they are unable to implement successful and sustainable measures in forest management in specific areas. In addition, a large number of forest-dependent people do not only rely on forests for subsistence or livelihoods, but forests are also seen as important for their culture and traditions. Depriving them of forests would consequently result in the loss of unique cultures, traditions and local knowledge. As forests play different roles in each community, Arnold (2001) noted, in order to understand what form social forestry might take in a particular situation it is therefore important to first identify how forests and forest outputs contribute to local livelihoods, how supply and use patterns are changing, and the reasons for, and consequences of, these changes.

The importance of forests and forest products to rural residents has been increasingly recognized. Since the 1990s, indigenous people have received considerable attention regarding their involvement in community-based forest management, based on the fundamental rights of these communities to exercise control over their natural resources (Gilmour, 2016) (see Table 1).

In Indonesia, as in many parts of the world, indigenous groups have struggled for legal recognition of traditional (often considered as informal) land rights claims, despite having over time developed some sort of institutional arrangements to govern the use of local natural resources (Gilmour, 2016). Many indigenous peoples have created sustainable systems of forest use based on their knowledge of forest and forest products, often referred to as indigenous knowledge (Arnold, 2001). A report published by World Resources Institute (WRI) in 2016 on an economic cost and benefit analysis of tenure security in indigenous forestlands in Bolivia, Brazil, and Colombia stated that tenure-secure indigenous forestlands exhibit low deforestation rates and that securing indigenous forestland tenure is a low-cost high benefit investment and has significant potential for cost-effective carbon mitigation., Recent research has also suggested that incorporation of relevant indigenous technical knowledge in community forestry policy formulation and implementation is needed to achieve effective community-based forestry systems (Gilmour, 2016).

The concept of social forestry and community-based forest management emerged in Indonesia in the late 1990s, following the fall of the Suharto regime, also known as the New Order regime (Siscawati et al., 2017). During the Suharto regime, natural resources, including forest resources, were viewed as the main sources of economic growth, and state control over forest areas was emphasized. It was discovered that by 1990, more than 500 companies had received concession licenses from the government forest authority, and approximately 60 million ha of forest area was offered for logging concessions (Yasmi et al., 2005 in Siscawati et al., 2017). This production system undoubtedly fostered an explosive increase in land and forest conversion. Moreover, since the mapping of forest area neglected local and customary land uses, many areas that were granted by the Ministry of Forestry to companies actually belonged to customary territories and were home to many local communities (Siscawati et al., 2017). With an absence of legal proof of ownership, local communities suffered as their rights and access to forest were restricted or removed. As a result, conflicts occurred due to overlapping claims over land. In order to tackle the conflicts, government agencies and companies often opted for coercive measures, including the use of the military to uproot the communities (Siscawati et al., 2017).

In response to this, several customary communities along with other local communities in Indonesia began to organize a resistance movement. They formed alliances with NGOs, academia and experts, and worked together under the Community Forestry Communication Forum (FKKM) (Siscawati et al., 2017). They aimed at pressing for forest tenure reform and the revision of the Basic Forestry Law by providing an alternative draft of a new forestry law to the Indonesian parliament just a few months before the New Order regime collapsed (Siscawati et al., 2017). In response to the pressure and growing awareness, some concessionary changes were initiated by government agencies. Change in the form of government, from a centralized to a decentralized one, was also seen as a positive influence, as it gave hope of greater autonomy for provincial and district governments. Accordingly, in 1999, the initial government of the post-New Order era enacted the Forestry Law No. 41/1999. The law serves as the legal basis for forest management, including social forestry schemes, in Indonesia. In 2008, the government revised and enacted more detailed regulations, PP No. 3/2008, covering the mechanism for issuing licenses for social forestry schemes.

Based on Forestry Law No. 41/1999, there are two forest tenure categories, state forest and private forest (See Fig. 1). Although the law recognizes the existence of customary forest, it initially categorized customary forest as a part of state forest, not private forest. This means that the management right of customary forest belongs only to the government, not to the local community. Alluding to this fact, the Alliance

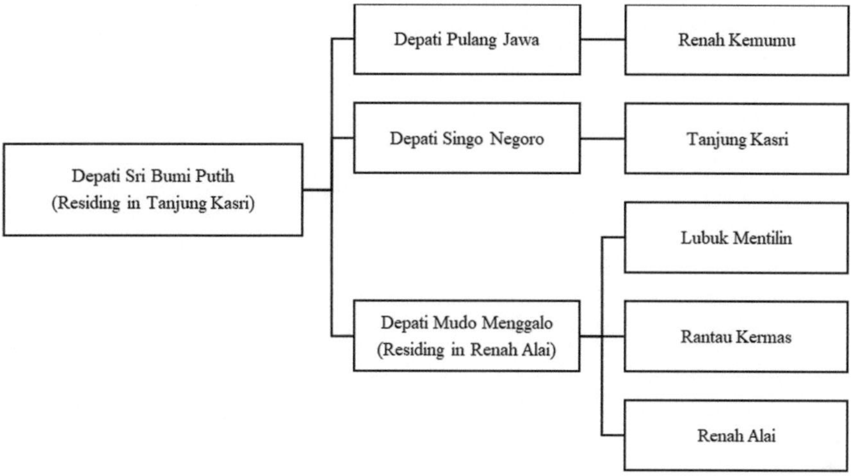

Fig. 2 Leadership hierarchy in marga serampas community

of Indigenous Peoples of the Archipelago (AMAN) argued that Forestry Law No. 41/1999 did not recognize the independent land rights of indigenous peoples (Siscawati et al., 2017). AMAN later submitted an official request for a judicial review of the law on March 19, 2012, based on the fact that the law contradicted provisions of the Constitution of Indonesia, as the 2002 Amendment to the Indonesian Constitution does recognize the cultural identity and traditional rights of indigenous peoples as basic human rights (Siscawati et al., 2017). Subsequently, on May 16, 2013, the constitutional court decided that customary forests were no longer part of state forests, and redefined the status of customary forest as private forest (See Fig. 2). In December 2016, four years after the ruling, the government issued customary forest permits for nine customary communities in Indonesia, including Marga Serampas. The following year, 2017, the government granted nine more permits. (see Table 2).

Social forestry programs also possess several challenges as well as limitations in their implementation. Anderson et al. (2015) noted that technical, managerial, and financial requirements stipulated by community forestry or participatory forest management frameworks may be incompatible with local realities and interests. In many cases, despite policy reforms that ceded direct control of valuable resources to local communities and local governments, performance was constrained by poor local resource management capabilities. Baynes et al. (2015) identified four central factors which influence the success of community forestry:

1. Power over resources are rarely voluntarily shared. Careful planning and implementation are required to mitigate conflicts and problems such as gender inequality, corruption and resource sharing injustice that may affect community cohesion. 2. Many governments pay lip-service to community forestry but continue to reduce budget allocations for its implementation, even though building a sustainable social forestry system will likely require long-term capacity building. 3. Both short-term

Table 2 Customary forests and customary communities (*Masyarakat hutan adat*) in Indonesia with legal permits as of October 2017

No	Name of customary forest/customary community	District	Province
1	Marga Serampas	Merangin	Jambi
2	Bukit Sembahyang & Padun Gelanggang	Kerinci	Jambi
3	Bukit Tinggai	Kerinci	Jambi
4	Tigo Luhah Permenti Yang Berenam	Kerinci	Jambi
5	Tigo Luhah Permenti Kemantan	Kerinci	Jambi
6	Ammatoa Kajang	Bulukumba	South Sulawesi
7	Wana Posangke	North Morowali	Central Sulawesi
8	Kasepuhan Karang	Lebak	Banten
9	Tombak Haminjon (Kemenyan)	Humbang Sundutan	North Sumatra
10	Tapang Sambas-Tapang Kemayau	Sekadadu	West Kalimantan
11	Kulawi	Sigi	Central Sulawesi
12	Dusun Senamat Ulu	Bungo	Jambi
13	Kampung Juaq Asa	West Kutai	East Kalimantan
14	Dusun Baru Pelepat	Bungo	Jambi
15	Desa Ngaol	Merangin	Jambi
16	Desa Pulau Tengah	Merangin	Jambi
17	Dusun Batu Kerbau*	Bungo	Jambi
18	Dusun Batu Kerbau*	Bungo	Jambi

*: Consists of two different communities

cash income and long term material benefits may be needed if forests are to be managed sustainably. 4. Lastly, perhaps the most complex factor determining the success of community forestry relates to the treatment of land and tree tenure Tenure security highly affects local people's motivation. Research on village forest (*hutan desa*) management in Sumatra and Kalimantan by Santika et al. (2017) found that the management has significantly reduced deforestation, despite varying performance in each village. Similarly, the research suggests that capacity building for community members, adequate assistance and financing from the government, and partnership with external factors are imperative to build a strong foundation for the management of customary forest. Anthropogenic pressures influenced by economic motives, especially in areas with intense anthropogenic pressure (e.g. low lands, established agriculture land, areas close to major roads) may become a main challenge in the future (Santika et al., 2017). In addition, issues related to the land tenure system, such as overlapping claims for state forests and customary forest, and significant information gaps among different actors and institutions regarding the land and forest tenure system in Indonesia can pose a challenge to effective implementation (Siscawati et al, 2017). Siscawati et al. (2017) also pointed out that the process of obtaining a social forestry permit is too complicated and technically difficult for local communities. With at least 26 necessary steps, the process can take up to three years to complete. Yet actors involved in land tenure reform in Indonesia seem to have limited information and knowledge on how to simplify the process. The capacity of local communities for claiming, enforcing or defending their traditional rights has also been a challenge, since many of them have limited awareness of the existing regulations and are not equipped with adequate technical knowledge (e.g. preparing maps, management plans, etc.) (Siscawati et al., 2017).

Taking these points into account, in the next section we will present a case study of social forestry in Rantau Kermas Village in Sumatra and discuss the issues and challenges faced in implementation.

3 Overview of Forest Conditions and Forest-Dependent Communities in Sumatra

The island of Sumatra, located in western Indonesia, is one of the biggest islands in the country comprising an area of 473,500 Km^2 – roughly twice the size of Honshu island, the biggest island in Japan. The island is endowed with diverse rainforests and is home to many unique and endangered species. The island plays an important role to the country's economy as it contains a large number of oil palm, pulp, and forest timber plantations, with Riau claiming the largest number. Sumatra is also known as the largest producer of coffee. The history of monoculture cash crop production on the island dates back to Dutch colonial times (Kunz et al., 2017). In fact, the current government still perceives Sumatra as a center for resource production and key to the nation's strategy for economic development and poverty alleviation (Kunz et al., 2017).

This intensive agricultural conversion for production on a massive scale in recent years, accompanied by increasing population pressures, has led Sumatra to experiencing an alarming rate of deforestation and forest degradation. By 2010 Sumatra was estimated to have converted 70 percent of the island's total forested area for various purposes, mainly agricultural expansion and establishment of plantations (Margono et al., 2012). Margono et al. (2012) found that between 1990–2010, Sumatra was estimated to have experienced a total of 7.54 Mha of forest loss, of which 7.25 Mha was in a degraded state when cleared and 0.28 Mha was in a primary state. The study also found that the rate of forest cover change slowed over this period from 7.43 Mha in 1990–2000 to 2.51 Mha in 2000–2010 (Margono et al., 2012). Therefore, in response to the intensive exploitation of its natural resources, Sumatra is facing a huge challenge to be able to simultaneously sustain key ecosystem services in its forests as well as the livelihoods of forest-dependent people.

The province of Jambi is located in the central part of Sumatra, neighboring Riau province. It comprises a population of 3,406,178 people and has an area of 50,058.16 Km^2. In 2013, forest land in Jambi covered a total area of 21,077.79 Km^2, which is equal to more than 40 percent of the total land area of the province (INCAS, 2015). Following Riau, Jambi has the second highest production of palm oil among Sumatran provinces, with an estimated total area of 941,000 ha (Kunz et al., 2017). However, despite the significance of palm oil production, rubber remains the dominant crop in Jambi, covering an area of approximately 1,284,000 ha (Gatto et al., 2015; Kuntz, 2017). Other widely produced crops in Jambi include cinnamon and coffee.

Jambi is a culturally and ethnically diverse province. According to the Ministry of Home Affairs, the original inhabitants of Jambi province consist of various ethnic groups including Malay Jambi or *Melayu,* Batin, Kerinci, Penghulu, Pindah, Anak Dalam (also known as Kubu), and Bajau. The Melayu ethnic group makes up the majority of population and is often referred to as 'autochthonous' by researchers. They originally settled along the Batanghari river. Nowadays, Jambi serves as a melting pot of people from across Sumatra as well as migrants from other islands such as Java.

In spite of the generally increasing heterogeneity found in urban areas and cultural assimilation among the people, several ethnic groups in Jambi still follow their old traditions and customs, or *adat* as it is called in Indonesian language. Some of the members of these ethnic groups live a rather secluded and culturally distinctive lifestyle in rural parts of Jambi, and settle in areas adjacent to forest. The majority of ethnic group members living near forest areas also rely on forests for their livelihoods. However, customary communities in Jambi, as elsewhere in Indonesia, are often marginalized and accused of encroaching on the forest (Nugroho, et al., 2017). This is especially the case when they are unable to prove the legality of their de jure land rights. Most of them employ traditional land-use systems that have been implemented for many generations yet are considered informal by the government. This overlapping land tenure, both formal and informal, is often confusing and makes land conflict resolution extremely complicated. Recent rampant agricultural expansion, accompanied by the lack of land tenure security, consequently puts these people at risk when conflicts arise between interest groups. Social forestry has thus gained currency as a potential solution. Indeed, as of the end of 2017, there were eighteen customary communities (*masyarakat hukum adat*) nationwide that have been granted legal recognition of their ownership over customary forests, and eleven of them are located in Jambi (see Table 2). The significantly high number of forest recognition in Jambi is largely attributed to the long-term involvement of NGOs in organizing local communities and the supportive attitudes of district and provincial governments towards social forestry policies (Bock, 2012; Urano, 2014). The establishment of customary forest (*hutan adat*), as well as other social forestry schemes in Indonesia, involves an arduous process that requires extensive facilitation and community participation, which might be difficult without vigorous support from NGOs. Despite their varying interests, NGOs in Jambi have collaborated to develop a unified regional approach to support communities in securing long-term forest management rights through multiple approaches such as advocacy, capacity building, legal support, generating awareness, and providing services (Bock, 2012).

4 Case Study: Customary Forest (*Hutan Adat*) in Rantau Kermas Village

Rantau Kermas Village is situated in a mountainous area in Jangkat district, Merangin regency, Jambi province, on the border with Bengkulu province. The village is one of five adjacent villages comprising the Marga Serampas customary community (the others are Renah Alai, Lubuk Mentilin, Tanjung Kasri and Renah Kemumu). It takes approximately eight to ten hours to drive from Jambi city to the village and it is 145 km away from the closest city, Bangko, the capital of Merangin regency. When the first author visited the village during the rainy season in late November 2017, cars were not able to negotiate the muddy, steep and potholed road leading to Rantau Kermas from the neighboring Pulau Tengah village, so only motorcycles could be used.

Rantau Kermas comprises an area of 67 Km^2 that makes up 6.93 percent of the total land area of Jangkat district (BPS-Statistics of Merangin Regency, 2017). According to statistics of Merangin Regency, the population of the village in 2016 was 422, with 212 males and 210 females. There are about 98 households in the village and each household has an average of 4–5 people. According to the village leaders, Rantau Kermas was designated as a Kampung Keluarga Berencana (*family planning village*) five years ago. Since then, they have received workshops and ounselling about family planning programs in which residents are encouraged to use contraception to limit their children to only two. Everyone in the village is Muslim, following an orthodox Islamic tradition that is strongly incorporated into Serampas *adat* laws. Most are originally members of the Serampas customary community and born and raised in the area, although some of the residents originated in other parts of Indonesia but married into the Serampas community.

The village has one kindergarten and one elementary school. During our visit, we did not meet any teenagers from junior high school to high school age, as children in Rantau Kermas are sent to school or *pesantren* (Islamic boarding school) in Bangko or Jambi city upon graduating from elementary school. The children normally return to the village twice every year during school holidays and in the month of Ramadan for the Eid-ul Fitr celebration. Some of them continue higher education until university level, but many of them choose to return to the village after they graduate from high school. The major occupation for both men and women in the village is farming.

5 Serampas as a Customary Community and Territorial Entity

As mentioned earlier, Rantau Kermas is one out of five villages in the Marga Serampas customary community. The name, Serampas, refers both to the people belonging to the customary community as well as the region occupied by them (Hariyadi & Ticktin, 2012). The territory of Serampas is widely known by the majority of people in Jambi, even though their legal tenure is considered to be informal. It is said that the people of Serampas have inhabited the region since sometime in the eleventh to thirteenth century (Bonatz et al., 2006). The members of the community are well-known for following *adat* or customary laws.

Masyarakat adat is a distinct term for describing customary communities in Indonesia. The term is also often used to refer to the indigenous peoples of Indonesia, since the country does not have any legal definitions or special classifications of indigenous peoples, although there is debate as to whether the terms indigenous and *adat* can be used interchangeably. Some researchers argue that the international legal category of 'indigenous peoples' may not be appropriate to describe Indonesia's situation (Kusumaatmadja, 1993 in Li, 2001). It is said that either all the people native to the archipelago are indigenous or no one is (Kusumaatmadja, 1993 in Li, 2001). The basis for this argument, according to Li (2001), is the fact that in Indonesia, there is no history of reserving land for particular ethnicities or groups or using race to understand differences between groups. Indonesian people normally use the term *suku* to refer to particular ethnic groups. During the Suharto regime, the dominant Javanese were accused of ethnocentrism in their superior attitudes towards the residents of the outer islands, especially rural villagers (Dove, 1985 in Li, 2001). In that era, the lifestyle of rural villagers was seen as obsolete and not befitting the ideal image of 'modernized and economically developed' Indonesia. Hence, the denigration of rural peoples in the outer islands occurred that in some instances still prevails today. AMAN (Aliansi Masyarakat Adat Nusantara), the Alliance of Indigenous Peoples of the Archipelago, describes indigenous peoples as 'communities which have ancestral lands in certain geographic locations and their own value systems, ideologies, economies, politics, cultures and societies in their respective homelands.' Similar kinds of definitions are used widely in Indonesia. As a matter of fact, the majority of Indonesians lay claim to a particular cultural heritage rooted in a certain region; thus they could declare themselves as 'indigenous.' Taking this into account, Marga Serampas does fit into the general, widely-used definition of indigenous peoples in Indonesia.

Members of Marga Serampas inhabit a territory that lies within Kerinci Seblat National Park. The national park itself spreads over four provinces, West Sumatra, Jambi, Bengkulu, and South Sumatra, but Marga Serampas community members settle specifically within the area in Jambi province. The massive national park area is home to many endangered animals such as the Sumatran tiger, tapir and elephants. More importantly, the national park encompasses source springs for the largest rivers in south Sumatra, the Batanghari, Musi and Merangin.

Hariyadi and Ticktin (2012) wrote that Marga Serampas members consider themselves to be of Minangkabau descent, or originally from West Sumatra. They concluded that the Marga Serampas community does share similar socioeconomic activities, dialect, and culture. Indeed, similar findings were also observed during the first author's visit to Rantau Kermas village. In interviews and focus group sessions, villagers of Rantau Kermas often mentioned that their *ninik-mamak*, the local word for ancestors, are from Minangkabau in West Sumatra and that many of the customary laws they continue to follow were adapted from the values and traditions shared among Minangkabau communities. For example, the inheritance of rice fields (*sawah*) in Marga Serampas is through the female line (*matrilineal*), just as with the Minangkabau. Inheritance of properties through the female line is not very common in Indonesia, and the Minangkabau is one of the few ethnicities that do so. In addition, members of Marga Serampas embrace Islam, as do the Minangkabau. They follow an orthodox Islamic teaching that is intertwined with their local and cultural traditions. The strong influence of Islamic values can also be found in most of their *adat* laws. Each Marga Serampas village has one *imam* who serves as a religious leader in the community. An *imam* plays an important role to ensure that the society complies with the mutually supporting Islamic and *adat* rules. He also serves as an advisor to a group of *adat* leaders to ensure that no ruling is against Islamic values. However, despite strongly adhering to Islamic values, many Marga Serampas residents are believed to practice black magic (*ilmu hitam*) and possess supernatural abilities (Neidel, 2006). This runs counter to orthodox Islamic teaching, which considers magic as *shirk,* or an act of polytheism which is largely considered sinful. When the first author traveled from Jambi city to Rantau Kermas village, many locals warned about the black magic along the way. Local residents advised being careful and cautious about not breaking *adat* rules or offending Serampas people, as they might cast black magic upon others. However, as Neidel (2006) pointed out, the practice of traditional black magic in the area is declining. In an interview with the *imam* of Rantau Kermas village he said that most people in the village now acknowledge that magic and certain rituals in *adat* are not part of Islamic teaching so should be eliminated from their daily practice. Yet several other residents reported that they still perform particular (non-Islamic adat) rituals on certain occasions.

The *adat* laws in Marga Serampas serve as the foundation of both ethical and legal judgments, as well as a reference point for social expectations (Abdullah, 1966 in Hariyadi & Ticktin, 2012). When asked to provide examples of *adat* regulations during interviews, most residents gave a wide range of answers, citing social norms, religious rules, and the land tenure system within the community. In addition to the village head (*kepala desa*), who is elected as a government official, the five villages within Marga Serampas have *adat* leaders (see Fig. 2) called *depati*. Today there are four *depatis* who are leading the community. One *depati* of the highest rank named Depati Sri Bumi Putih serves as the general leader of the community. Another *depati* leads Renah Kemumu village, one leads Tanjung Kasri village, and another *depati* leads the other three villages, including Rantau Kermas. In past days, the position of *depati* was inherited but today people of Serampas vote for the main *depati* as well as those of the respective villages. According to members of Rantau Kermas village, *depatis* are more powerful than the official village head in terms of governing the society based on *adat*. Many residents, especially the elders, also stated that members of Marga Serampas still strongly follow the rules of *adat* and are optimistic that the next generations will do so. Should a problem occurr in the community, members first bring it up to the village *depati*, who will then consult with the imam and the community elders. If it cannot be solved at the village level, then the problem will be brought to the general *depati* and solved as a community problem. Members in

Rantau Kermas village said that the *adat* rules they are following today are the ones that their *ninik-mamak* had followed and passed on from generation to generation. However, most of them also said that there are some relatively new rules and that they are open to create novel rules to accommodate the needs of new generations. It is quite hard to trace the origins of these regulations as they are usually passed on through oral communication in the form of short poems in the local dialect.

In relation to livelihoods and resource management, Marga Serampas originally practiced shifting cultivation by cutting and burning primary or secondary forests to develop swidden areas (Hariyadi & Ticktin, 2012), a method commonly practiced in many traditional communities in Southeast Asian countries. Hariyadi and Ticktin (2012) noted that the Marga Serampas community in general recognize three different types of forests, primary forest, secondary forest and customary forest. Customary forests have been mainly preserved to protect watershed areas and steep forested lands that are susceptible to landslides (*ngarai*).

Since the 1970s, Marga Serampas has included the cultivation of cinnamon (*kulit manis*) as part of their main economic activities (Hariyadi & Ticktin, 2012). Since then, cinnamon has become an important commodity for the community and the neighboring district, Kerinci. This shows that there has been a shift in cultivation from farming for subsistence purposes to producing cash crops. Interestingly, other areas in Merangin regency mainly produce palm oil and rubber as their main agricultural commodities, and cinnamon plays a very small role in contributing to the regency's income. In their research on cinnamon cultivation in Marga Serampas, Hariyadi and Ticktin (2012) mentioned that the land tenure system in Marga Serampas has gradually changed since cinnamon became widely cultivated, with land that was formerly largely common property now tending to change to privately-held property. Earlier generations of Marga Serampas do not recognize the practice of buying and selling land, as it was all common property. Nowadays, community members are allowed to sell and buy usage rights to cinnamon agroforests. During an interview, some Rantau Kermas villagers admitted this fact and mentioned that as cash has increasingly become a basic need for the members, one should not refuse to buy the land offered by others especially if they are urgently in need of money to cover their basic needs.

One of the most notable *adat* regulations in Marga Serampas related to resource management is their land tenure system. According to Hariyadi and Ticktin (2012), adat ensures an equal distribution of land for farming and housing for all members of Marga Serampas. When community members get married, local elders and adat leaders conduct a ritual called *tanah ajum arah,* in which they guide the new couples to suitable vacant lands to build houses and farm land. Each household has the right to use the land that they are allocated but they do not own it. *Tanah ajum arah* in general acts also as a landscape management regulation for the area. It guides people to use certain lands and discourages them from using prohibited land such as the customary forest or steep forested lands.

Marga Serampas basically allows exogamous marriage between members of each village under Serampas territory and other villages. Outsiders are also allowed to settle in Serampas after they engage in several customary indigenous rituals and practices. One exception is for *orang selatan* or the people originally from South Sumatra or Lampung province in southern Sumatra, who are not allowed to marry nor to settle in Serampas territory. The reason behind this, based on the interview in Rantau Kermas village, is profound differences in custom and attitudes that make them incompatible. However, most members were uncertain about the history of this prohibition.

6 Customary Forest in Rantau Kermas Village: Opportunities and Challenges

Rantau Kermas village was granted an ownership license for its customary forest under the national social forestry program in December 2016. The village was one of the first customary communities to receive legal recognition of their ancestral forest lands. The enabling regulations were included in the ministerial decree of MoEF, SK.6475/MENLHK-PSKL/KUM.1/12/2016. The decree stipulated that the community must maintain the functions of the forest, which in the case of Rantau Kermas was exclusively for conservation. Among the social forestry schemes designated by the Indonesian government, customary forest is the most distinctive in regards to its regulation and most complex in the conditions that must be conformed to by the community (See Table 3). Unlike other schemes that only recognize management rights for a limited time but can be renewed, a customary forest scheme allows full ownership over a forest land. However, designating customary forest rights in Indonesia is rather complicated. First, customary community *(adat)* in Indonesia is a contested term that does not seemingly conform to the international definition of indigenous peoples. Customary communities in Indonesia are commonly associated with communities that follow traditional customary laws, live in rural areas, and have resided in their territories for a long time. Recognition of customary forest is based on the fact that forests play as an integral part in the economic and cultural activities of customary communities and these communities possess local environmental knowledge that has allowed them to sustain their local resources. By incorporating this knowledge in the social forestry system, the government hopes that customary communities can contribute to reduce deforestation and land degradation by managing the forest in a sustainable way.

Before the recognition, members of Rantau Kermas village were involved in a dispute over the customary forest with officials from Kerinci Seblat National Park. National Park officials claimed that part of the forest is included in the park area and must be free of human activities, whereas the Rantau Kermas villagers as well as members of Marga Serampas strongly believe that the customary forest is part of their ancestral lands. Although Marga Serampas members are not allowed to cut down trees in customary forests, they are still allowed to take non-timber forest products (NTFPs) such as rattan for personal use. Another condition is that the customary forest land is not allowed to be sold. The total area of customary forest in Rantau Kermas

Table 3 Social forestry schemes and general regulations adapted from: Herawati, 2017

Category	Location	Permit	Applicant	Licensor/ Grantor of permit	Status and Period
Village forest	Production Forest & Protected Forest	HPHD (Village Forest Management Permit)	Village cooperatives/ enterprises	MoEF/ Governor	35 years and can be extended
Community Forestry	Production Forest & Protected Forest	IUPHKm	Community representatives	MoEF/ Governor	35 years and can be extended
Community Plantation Forest	Production Forest	IUPHHK-HTR	Individual/ Groups/ Cooperatives	MoEF/ Governor	35 years and can be extended
Collaborative forest management	Production Forest, Protected Forest & Conservation Forest	Agreement	MoU between concessioner and community groups	MoEF/ Governor	35 years and can be extended
Customary Forest	Customs Territory, outside state forest	Forests subject to rights/ Customary forest	Customary groups (Masyarakat Adat)	MoEF	Ownership rights

village is estimated to be 130 ha, of which 24 ha were considered to be part of Kerinci Seblat National Park before the decree. In obtaining the recognition, the community was assisted by a Jambi-based conservation NGO called KKI Warsi, one of the most prominent conservation organizations in Jambi. KKI Warsi has cultivated a strong relationship with the village since the early 2000s and played an important role in facilitating the process of obtaining the customary forest license for Rantau Kermas residents. Both community members and representatives of KKI WARSI stated that the process in receiving the license was complicated and time-consuming.

People of Rantau Kermas village mainly grow coffee and cinnamon as their commodities. They also grow rice but it is mainly for subsistence, not for sale. They have a rule called '*timpo menanam*' which which obliges young men who have completed their education (normally at the high school level) to grow at least 1000 coffee plants if he is unmarried and 2500 plants when he is married. These young men will be allocated a certain area of land and must grow the expected amount of coffee within a specific period of time. Many members said they find this rule provides them with a sense of security as they know that they will have a source of living after they finish school. Others mentioned that this rule creates a bond in the form of responsibility between young men and their community. They feel obliged to come back to the village after they complete their education, normally in cities such as Bangko or Jambi.

In managing their farm lands, the villagers employ a system called '*barin*' in which farm owners take turns in working in each other's farm land or rice fields. The work is done by both men and women. The owner of the land they are working on will be in charge of preparing lunch for the day. The farm work usually begins at 7 am and finishes at 4 pm. Rantau Kermas villagers still strictly follow this system. However, the villagers noted that residents of a neighboring village, Renah Alai, that is also more advanced in terms of infrastructure and accessibility, are slowly abandoning the *barin* system in favor of a paid labor system as their work force is declining in number.

As regards the customary forest, review of findings from participant observation, interviews, and a focus group session in Rantau Kermas village reveals that villagers are aware of the importance of preserving the customary forest since *adat* rules prohibit them from cutting down the trees or converting the forest into other purposes. Besides, they also know that the forest plays a significant role in providing them with sources of clean water as well as preventing landslides. Generally, the villagers showed a keen interest and strong willingness to protect their customary forest.

Several issues and challenges related to implementation and community performance in customary forest management were also observed. First, within the community itself, there is an issue of knowledge and information gaps between the older and younger generations. As with many customary communities in Indonesia, villagers in Rantau Kermas rely mainly if not solely on oral information passed on from one person to another. This was rather easy to maintain in former years due to limited access to communication, as there was no phone reception nor 24-h electricity in the village, or physical access such as roads to outer areas. Since most villagers stay in the village and perform most of their activities communally, transfer of knowledge can be done virtually anytime. The transfer of knowledge is important to ensure that that even the youngest generation at the village understand the local wisdom and customs of the community, including those related to forest and land management. Today a micro-hydropower plant is being renovated, which means that they will have access to electricity soon. With the introduction of electricity, the villagers' daily activities will surely change. They might prefer to stay at home and become less involved in social activities. With new media and improved access to the outside world the villagers might pick up behaviors or ideas that don't conform with *adat* rules. The teenagers of Rantau Kermas normally return to the village after they graduate from higher education; and learn how to farm afterward. Young men from the village are induced to come back because of the '*timpo tanam*' rule and the fact that they can be allocated land for a house and production. However, attending schools in urban areas exposes them to many alternative occupations besides farming. Although most of the parents were optimistic that they will come back and stay in the village, some also mentioned how they actually hoped the children could make a better living in the city rather than become farmers in the village. If the younger generations are not equipped with adequate knowledge of *adat* and prefer to leave, it will have negative impacts on the continuity of traditions, which are not only cultural but also related to forest and resource management. Fields, forests and plantations in the village might be abandoned, and their local wisdom and tradition will fade away. As the number of people who have the capability of managing the forest declines, the future generation might mismanage the forest or choose to convert their lands into something else or even sell it. In order to deal with this challenge, there is an urgent need for capacity building for forest-dependent people to ensure that they have the capabilities to protect their forests, both physically and in legal terms. Besides, as most forest communities have limited awareness regarding forest regulations and do not possess adequate technical ability, capacity building is also important for the communities to be able to claim, enforce and defend their rights.

Furthermore, easy access will also ease movement and migration. During the first author's visit, the road to Rantau Kermas village was in poor condition, but many villagers mentioned that there is a plan to renovate the road soon. Research by Angelsen and Kaimowitz in 1999 showed that greater access to forests and markets does accelerate deforestation. Certainly, despite the legal license that it owns, the threat of land grabbing and land conversion still exists in Rantau Kermas. Jambi is located next to Riau province, with the highest rate of forest and peatland loss, followed by South Sumatra. The main cause for this is arguably palm oil plantations. With better roads and access, outsiders can easily come and enter the Rantau Kermas area. This might put the customary forest in a vulnerable position. Besides, Rantau Kermas villagers generally allow outsiders to freely enter the village. Yet without adequate knowledge and awareness, villagers might be tempted or tricked to sell their lands to outsiders.

Other challenges are related to external factors. Success in obtaining a license has largely been attributed to the facilitation provided by KKI Warsi. KKI Warsi not only supported the community through advocacy but also provided them with technical assistance (e.g. mapping, educating the villagers about the legal system) during the application process and implementation of the policy. The arduous process of application and complex requirements make it almost impossible for forest communities in Indonesia to be independent in achieving the permit and implementing the program. There is a need to simplify the process and increase government support. Otherwise, forest-dependent communities would have to establish and maintain a good long-term relationship with a facilitating NGO. In addition, monitoring and assessing the progress of customary forest, as well as other social forestry schemes, can be challenging for the government. Similar to Rantau Kermas, many customary forests in Indonesia are located in remote areas, so the government must set aside adequate budget and human resources for monitoring and financing the social forestry programs. Indeed, the challenges that were observed in Rantau Kermas village are comparable to the challenges described by other researchers in the previous section. Lastly, since the social forestry program in Indonesia is relatively new, the government should prescribe certain criteria to evaluate the success of program and what factors contribute to it. This is important to be able to measure the capability of the community and other supporting institutions in ensuring the long-term sustainability of the program. Aside from that, the assessment will be helpful for identifying the benefits and challenges of implementing social forestry programs in different communities. Jambi is one of the leading provinces in terms of the social forestry program. However, there are still many other locations that require attention. As

Indonesia consists of highly diverse forest-dependent communities with multiple needs and concerns, assessing the performance of existing social forestry programs will help achieve better performance in the future and keep community members as other stakeholders informed about their current situation.

7 Conclusion and Recommendations

In conclusion, the social forestry program is perceived as promising by some and as little more than a trendy movement that may not be very sustainable in the long run by others. In Indonesia, the implementation of social forestry program is still in its early stage. Despite being declared by many scientists and policymakers as a win–win solution for both forests and communities to reduce deforestation, forest degradation and land conflicts, many questions, issues and challenges remain, especially regarding how the program is implemented and the benefits it brings to forests and local communities.

A case study of customary forest in Rantau Kermas village shows that there are several challenges and issues that arise in the implementation of a social forestry program in a customary community. First of all, deciding which community can be categorized as an *adat* community is still highly debatable. Since there is no single definition presented by the government, it opens a possibility for any group to claim themselves as an *adat* community. This can be problematic since if a community is granted customary forest recognition, it will have to ensure the protection of its customary forest as long as the community continues to exist. If the ownership of forests fall into the wrong hands, ensuring the protection of forest might be questionable. Besides, although Jambi province is leading in terms of the number of customary forests, there are still many other *adat* communities in outer islands that are waiting to be granted their rights. The process of legal recognition for an adat community can be complicated, requiring assistance from a third party such as an NGO to assist indigenous people who may not be well-informed about state regulations and procedures. Once the community is granted its rights over forests, questions may arise about the government's and community's preparedness to ensure the sustainability of the program. On top of that, whether or not the future generations will be willing to maintain the existence of *adat* and follow the traditional customary rules along with the ever-changing dynamics in society will surely have impacts on the sustainability of the program. Lastly, another challenge is understanding and determining the benefits of the customary forest scheme to the wellbeing and livelihoods of the community and the environment. The government should be able to assess the performance of each community and be informed about the benefits of a particular social forestry scheme in each area in order to form a sustainable forest management plan that suits multiple needs of the highly diverse communities in the future.

As for recommendations, if similar research is to be done in the future, it would be beneficial to look at the process behind customary forest recognition to see what factors might contribute to improved implementation. In the case of Rantau Kermas and Jambi, it seems that NGOs play an active role in advocating for customary communities to the government. They can also be active in educating the locals about relevant issues. Considering that land disputes still occur and that there are still many customary communities in remote rural areas in Indonesia that may benefit from the recognition of their forests, research on the process of recognition may help them in identifying the key factors that might bring their communities to success. Aside from that, more research on the challenges of implementing customary forestry and the social forestry program in Indonesia is needed. Since the program itself is fairly new, identifying the challenges and issues that arise will contribute to the improvement of the program itself. Besides, it is important to ensure the continuity and the sustainability of the program. Involving local people in forestry activity sounds feasible when the local communities have their own sustainable ways in managing their forest resources. However, changes within the community may happen along with improved infrastructure, education, and migration. These changes may impact the way communities use and manage their resources, as well as the long-term existence or identity of the community itself.

References

Anderson, J., Mehta, S., Epelu, E., & Cohen, B. (2015). Managing leftovers: Does community forestry increase secure and equitable access to valuable resources for the rural poor? *Forest Policy and Economics, 58*, 47–55. https://doi.org/10.1016/j.forpol.2014.12.004

Angelsen, A., & Kaimowitz, D. (1999). Rethinking the causes of deforestation: Lessons from economic models. *The World Bank Research Observer, 14*(1), 73–98.

Arnold, J. E. M. (2001). *Community forestry*. Food and Agriculture Organization of the United Nations.

Banjade, M. R., Herawati, T., Liswanti, N., & Mwangi, E. (2016). Forest tenure reform in Indonesia: When? What? Why? *Center for International Forestry Research*. https://doi.org/10.17528/cifor/006333

Baynes, J., Herbohn, J., Smith, C., Fisher, R., & Bray, D. (2015). Key factors which influence the success of community forestry in developing countries. *Global Environmental Change, 35*, 226–238. https://doi.org/10.1016/j.gloenvcha.2015.09.011

Bock, M. J. (2012). Formalization and Community Forestry in Jambi, Indonesia: Indigenous Rights, rural migrants, and the informal divide.

Bonatz, D., Neidel, J. D., & Tjoa-Bonatz, M. L. (2006). The megalithic complex of highland Jambi: An archaeological perspective. *Bijdragen Tot De Taal-, Land-En Volkenkunde/journal of the Humanities and Social Sciences of Southeast Asia, 162*(4), 490–522.

BPS-Statistics of Merangin Regency. (2017). *Kecamatan Jangkat Dalam Angka 2017 (Jangkat Subdistrict in Fig.s 2017)*. Jambi, Indonesia: BPS-Statistics of Merangin Regency. Retrieved from https://meranginkab.bps.go.id/adminweb/pdf_publikasi/Kecamatan-Jangkat-Dalam-Angka-2017.pdf

Carlson, K. M., Curran, L. M., Ratnasari, D., Pittman, A. M., Soares-Filho, B. S., Asner, G. P., & Rodrigues, H. O. (2012). Committed carbon emissions, deforestation, and community land conversion from oil palm plantation expansion in West Kalimantan, Indonesia. *Proceedings of the National Academy of Sciences, 109*(19), 7559–7564. https://doi.org/10.1073/pnas.1200452109

Food and Agriculture Organization of the United Nations. (2015). *Global forest resources assessment.*
FWI. (2014). *Potret Keadaan Hutan Indonesia Periode 2009–2013*. Bogor, Indonesia: Forest Watch Indonesia. Retrieved from http://fwi.or.id/wp-content/uploads/2015/05/PKHI-2009-2013_update__sz.pdf
FWI/GFW. (2002). *Potret keadaan hutan Indonesia*. Bogor, Indonesia : Washington, D.C., U.S.A: Forest Watch Indonesia ; Global Forest Watch.
Gatto, M., Wollni, M., & Qaim, M. (2015). Oil palm boom and land-use dynamics in Indonesia: The role of policies and socioeconomic factors. *Land Use Policy, 46*, 292–303. https://doi.org/10.1016/j.landusepol.2015.03.001
Geist, H. J., & Lambin, E. F. (2002). Proximate causes and underlying driving forces of tropical deforestation. *BioScience, 52*(2), 143. https://doi.org/10.1641/0006-3568(2002)052[0143:PCAUDF]2.0.CO;2
Gilmour, D. A. (2016). Forty years of community-based forestry: a review of its extent and effectiveness. *Food and Agriculture Organizations.*
Hariyadi, B., & Ticktin, T. (2012). From shifting cultivation to cinnamon agroforestry: changing agricultural practices among the serampas in the Kerinci seblat national park. *Indonesia. Human Ecology, 40*(2), 315–325. https://doi.org/10.1007/s10745-012-9481-8
Herawati, T. (2017). Brief of community forestry regulation in Indonesia [Powerpoint slides].
Hosonuma, N., Herold, M., De Sy, V., De Fries, R. S., Brockhaus, M., Verchot, L., & Romijn, E. (2012). An assessment of deforestation and forest degradation drivers in developing countries. *Environmental Research Letters, 7*(4), Article 044009. https://doi.org/10.1088/1748-9326/7/4/044009
Koh, L. P., & Ghazoul, J. (2010). Spatially explicit scenario analysis for reconciling agricultural expansion, forest protection, and carbon conservation in Indonesia. *Proceedings of the National Academy of Sciences, 107*(24), 11140–11144. https://doi.org/10.1073/pnas.1000530107
Kunz, Y., Steinebach, S., Dittrich, C., Hauser-Schäublin, B., Rosyani, I., Soetarto, E., & Faust, H. (2017). 'The fridge in the forest': Historical trajectories of land tenure regulations fostering landscape transformation in Jambi Province, Sumatra, Indonesia. *Forest Policy and Economics, 81*, 1–9. https://doi.org/10.1016/j.forpol.2017.04.005
Li, T. M. (2001). Masyarakat Adat, Difference, and the Limits of Recognition in Indonesia's Forest Zone. *Modern Asian Studies, 35*(3), 645–676.
Margono, B. A., Turubanova, S., Zhuravleva, I., Potapov, P., Tyukavina, A., Baccini, A., & Hansen, M. C. (2012). Mapping and monitoring deforestation and forest degradation in Sumatra (Indonesia) using Landsat time series data sets from 1990 to 2010. *Environmental Research Letters, 7*(3), Article 034010. https://doi.org/10.1088/1748-9326/7/3/034010
Molnar, A., Butterfield, R., Chapela, F., Fuge, P., De Freitas, A., Hayward, J., & Martin, A. (2004). Forest certification and communities. *International Forestry Review, 6*(2), 173–180.
Neidel, J. D. (2014). Discourse of decline: local perspectives on magic in highland Jambi. *Indonesia. Social Analysis, 58*(1), 67–87. https://doi.org/10.3167/sa.2014.580104
Nugroho, H. Y. S. H., van der Veen, A., Skidmore, A. K., & Hussin, Y. A. (2017). Expansion of traditional land-use and deforestation: a case study of an adat forest in the Kandilo Subwatershed, East Kalimantan, Indonesia. *Journal of Forestry Research.* https://doi.org/10.1007/s11676-017-0449-9
Provinsi—Kementerian Dalam Negeri—Republik Indonesia. (n.d.). Retrieved January 21, 2018, from http://www.kemendagri.go.id/pages/profil-daerah/provinsi/detail/15/jambi
Santika, T., Meijaard, E., Budiharta, S., Law, E. A., Kusworo, A., Hutabarat, J. A., & Wilson, K. A. (2017). Community forest management in Indonesia: Avoided deforestation in the context of anthropogenic and climate complexities. *Global Environmental Change, 46*, 60–71. https://doi.org/10.1016/j.gloenvcha.2017.08.002
Scherr, S. J., White, A., & Kaimowitz, D. (2004). A new agenda for forest conservation and poverty reduction: making forest markets work for low-income producers. Washington, D.C: Forest Trends.

Siscawati, M., Banjade, M. R., Liswanti, N., Herawati, T., Mwangi, E., Wulandari, E., … Silaya, T. (2017). *Overview of forest tenure reforms in Indonesia* (Working paper No. 223) (p. 45). Bogor, Indonesia: Center for International Forestry Research (CIFOR). https://doi.org/10.17528/cifor/006402

Tacconi, L. (2003). Fires in Indonesia: causes, costs and policy implications. CIFOR, Bogor, Indonesia.

Tarigan, S. D., Sunarti, & Widyaliza, S. (2015). Expansion of oil palm plantations and forest cover changes in bungo and merangin districts, jambi province, Indonesia. *Procedia Environmental Sciences, 24*, 199–205. https://doi.org/10.1016/j.proenv.2015.03.026

Tsujino, R., Yumoto, T., Kitamura, S., Djamaluddin, I., & Darnaedi, D. (2016). History of forest loss and degradation in Indonesia. *Land Use Policy, 57*, 335–347. https://doi.org/10.1016/j.landusepol.2016.05.034

United Nations Environment Programme (UNEP), & GRID--Arendal (Eds.). (2009). *Vital forest graphics*. Nairobi, Kenya: UNEP.

Urano, M. (2014). Impacts of newly liberalised policies on customary land rights of forest-dwelling populations: A case study from East Kalimantan. *Indonesia. Asia Pacific Viewpoint, 55*(1), 6–23. https://doi.org/10.1111/apv.12042

White, A., & Martin, A. (2002). *Who owns the world's forests? forest tenure and public forests in transition*. Forest Trends.

WRI. (2016). *Climate Benefits Tenure Costs*. World Resources Institue. Retrieved from https://www.wri.org/sites/default/files/Climate_Benefits_Tenure_Costs.pdf.

Sustainable Tourism in Japan: Kuriyama in Nikko City

Akiko Iizuka

Abstract The notion of sustainable tourism has been widely recognized since the 1990s. This chapter introduces the case of Kuriyama, Nikko City in Japan, which has been developed in rich culture and history. While Kuriyama has magnificent landscapes every season, challenges include the lessening of the labor force, natural disasters, poor accessibility and the merge into Nikko City that hinder local tourism to become sustainable. Several initiatives were taken by the local people and outsiders. The lessons learned from the initiatives present both tangible and intangible utilization of local resources, effective marketing strategies, the partnership within the community and the paradigm shift from limitation to potential.

Keywords Sustainable tourism · Community-based tourism · Rural revitalization · Local resource utilization · Cultural heritage · Tourism marketing

1 Introduction

Following the sustainable development stated in Brundtland report in 1987, the notion of 'sustainable tourism' has been widely recognized since the 1990s. Sustainable tourism is a long term growth in a specific area without deteriorating the quality and changing the environment (Butler, 1999). In order to promote sustainable tourism, local participation in the decision making process is necessary with the respect of local culture, history and tradition (Shimakawa, 2009, 30–31). The new approach is called 'landing-type' traveling commodities which put the people at the 'landing' spots centered in organizing and managing the travels (Okoso, 2016, 21–22). However, not all the local initiatives have gained positive results. In Japan, mass tourism based on packaged tours organized by travel agencies reached an impasse on the growth and the demand of domestic travels in Japan has dropped since the 1990s. While the population decreases in Japan, the demand of travelers has changed. With the development of information and communication technology, the demand

A. Iizuka (✉)
Center for International Exchange, Utsunomiya University, Utsunomiya, Japan
e-mail: iizuka@cc.utsunomiya-u.ac.jp

M. Yoshizumi et al. (eds.), *Navigating Local Sustainability in Food, Community, and Innovation*, https://doi.org/10.1007/978-981-96-8963-7_7

is diversified and thus new marketing systems are required. This chapter deals with the sustainable tourism in Japan exploring a case from Kuriyama, Nikko City. Can sustainable tourism be an alternative of mass tourism in Japan? Can sustainable tourism vitalize a community? What are the potential and limitations of the community? Most importantly, what lessons are learned from the case in order to achieve a better sustainable community in Japan?

2 Background of the Community

2.1 Potential of Tourism in the Community

Here is an illustration of Kuriyama from the Nikko City Tourism Association:

This area is known as the one of the most secluded regions that has a legend that fleeing Heike warriors who had been defeated in the Genpei War inhabited in the twelfth century. This area is also famous for undeveloped nature and the secluded hot spring, offering relaxed time when visitors can forget noisiness in the urban life (2018).

Kuriyama village was built by defeated soldier of *Heike* (Taira Families) to hide themselves about 800 years ago. Some members of *Heike* discovered the hot spring which gushed forth on the river bank and healed their injuries. The hot springs in Yunishigawa, Kawamata and Okukinu villages were developed since then and relaxing at hot spring resorts has been a very popular travel style in Japan. There are twenty Japanese inns (as hot spring resorts) among mountains at a height of 700 m. Kuriyama has been developed in rich history and natural resources therefore attracting tourists.

Kuriyama is located in Nikko City, the northwestern part of Tochigi Prefecture where Fukushima is in the north and Gunma is in the west (Fig. 1). New Nikko City was born in 2006 after a merge of five cities and towns including Imaichi City, Nikko City, Fujihara machi, Ashio machi and Kuriyama machi. Since the merge, Kuriyama has been one of the areas of Nikko City.

In addition to the rich culture and history of Kuriyama in Nikko City, the most attractive feature of Kuriyama is the natural landscapes from season to season. Kuriyama is located at the most north of Kanto area at a height of 700 m, therefore the landscape is enjoyable in every season; for example, summer resorts, colored leaves in autumn and deep snow during winter are especially famous among tourists.

In parallel with the seasonal changes, the cultural events are held every year. Table 1 show the events throughout the year which were handed down from a long time. For instance, *Heike taisai* is a festival for three days in the first weekend of June. It is to bring back the prosperity of Taira Family in thirteenth century into existence again and approximately 200 hundred people dress as warriors or princesses of the ancient dynasty and parade for three kilometers. It has been held for over 30 years and both local people and tourists join the event. From the middle of August to early

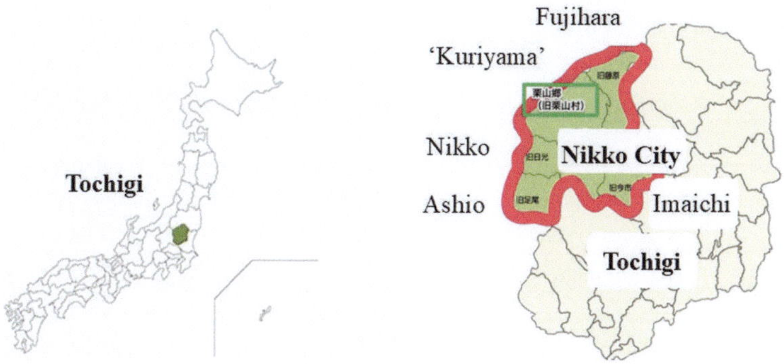

Fig. 1 Map of Kuriyama in Tochigi prefecture. *Source* Aoyama, 2017 (modified by the author)

September and October, the festival of *shishimai* (traditional lion dance) is held at all nine areas in Kuriyama. It has been a local performing art designated as an Important Intangible Folk Cultural Property of Nikko City and performed for good harvest and against misfortune. More importantly, the festival is the day when all the people including residents and relatives who live out of Kuriyama to get together in each area of Kuriyama. Since there is no high school and higher educational institutions, the young go out of Kuriyama from high school and mostly continue their study or work in the city.

In this way, almost every month, cultural and traditional events are held taking the advantages of the rich culture and the diversity of the seasons. Previously, mostly only local people joined such events, but more outsiders found the interests in the events after spreading the information through Social Network Service. The natural landscape and culture is seen as an ordinary thing to be handed down by the local people, however it is an attractive feature for visitors.

2.2 *Limitations of Tourism in the Community*

In spite of the rich culture and natural environment, there are several factors which hinder the local tourism in Kuriyama. The population of Kuriyama is 1,265 with 651 households accounting for only 1.5% of the entire Nikko City. The land area of Kuriyama covers 427.37 square kilometers accounting for 30% of Nikko City (2016). Kuriyama has a population density of just three per square kilometer (Nikko City, 2016) and one of the lowest population densities in Japan. The ratio of people 65 years old and over against total population is 43% in Kuriyama, compared to 27.3%, the total ratio in Japan (Cabinet office, 2017, 2), hence Kuriyama faces a serious aging problem. The population has decreased since 1980 with half the population remaining since 2000. In this sense, Kuriyama has the typical characteristics of remote rural areas of Japan facing the issues of depopulation and aging. The lack of labor force is

Table 1 Annual events of Kuriyama

Month	Events	Participants
January	*Genpuku* ceremony (coming-of-age ceremony for a boy in old days)	Mainly locals
February	*Setsubun* ceremony (bean-scattered ceremony to celebrate the coming spring)	Both locals and tourists
	Kamakura festival (snow festival) (Jan. ~ Mar.)	Tourists
June	*Heike taisai* (*Heike* festival)	Both the locals and tourists
	Ishiyaki (traditional hot stone barbecue)	Tourists
July	Summer walk	Mainly tourists
	Summer snow festival	Mainly tourists
	Take no yoi matsuri (bamboo evening festival)	Both locals and tourists
August	Summer festivals with *shishimai* (traditional lion dance) at eight locations (~Sep.)	Mainly locals
	Aurora fantasy event	Both the locals and tourists
September	*Goshinboku sai* (sacred tree festival)	Both the locals and tourists
	Suiri jinja sai (water shrine festival)	Mainly locals
October	A festival with *shishimai* (traditional lion dance)	Mainly locals
	Shin soba festival (newly harvested buckwheat festival)	Both the locals and tourists
	Kuriyama *Marche* (market)	Both the locals and tourists

a serious problem that prevents the promotion of local industries, especially tourism in the third sector which requires intensive labor for the services. Also, in terms of tourism, the information offered by Social Network Services has recently given an enormous influence on the people outside, however many elders have limitations to implement new services and find difficulty in responding to the needs of diversified tourists. For traditional Japanese inns and shops, the children of the owners generally take over the business as the successor when the owners retire. However, with the discontinuation of successors, people have to close down their hotels and shops. At the same time, creating job opportunities is therefore necessary for the young population to settle in Kuriyama to vitalize the local economy in rural areas. Consequently, without young people, there are missed opportunities; such as the development of new ideas, jobs, and innovation.

The factor that prevents local tourism in Kuriyama is the accessibility from cities like Tokyo. It will take approximately three hours from Tokyo to Kuriyama by car and more time by public transportation such as the train or bus. Compared to the other hot spring resorts such as in Hakone and Izu, the accessibility is limited for the people in Tokyo. Also, the transportation cost is high from Tokyo area to Kuriyama. For instance, the round trip costs approximately 7,000 Japanese yen by express train, 8,000 yen by car for the highway toll and 6,000 yen by bus depending on the season. Therefore, the associated time and the cost are negative factors that discourage possible tourists from Tokyo.

The other limitation of Kuriyama as well as Nikko City is the two disasters that occurred recently. The Tohoku earthquake and tsunami hit the northern parts of Japan in March 11, 2011 causing damages to a nuclear plant in Fukushima. Tochigi including Nikko City is located south of Fukushima; therefore, the local industry, especially tourism was seriously influenced. While recovering from the damages of the 2011 disaster, another earthquake hit the border between Tochigi, Gunma and Fukushima prefectures and especially Okukinu and Kawamata were seriously affected on February 25 in 2013 (Shimotsuke Newspapers, 2013); avalanches hit walking trails and two hotels in Okukinu and one hotel was closed down. In fact, the number of tourists was approximately 439,000 in 2009 and 420,000 in 2010; however, it dropped to 286,000 in 2011. In 2012, the number increased up to 309,000, but again dropped to 292,000 in 2013 and 278,000 in 2014. Since 2015, the number has rapidly recovered up to 382,000, but still not yet fully recovered compared to the pre-disaster period (Nikko City, 2010, 2011, 2012, 2013, 2014, 2015, 2016).

As stated previously, Kuriyama was merged into New Nikko City in 2006 and the name of Kuriyama was not shown in the official address although Nikko still keeps its name after the merge. Therefore, the withdrawal of the name through the merge might be preventing the brand name of Kuriyama getting into the public eye because their name is not registered as an official address.

3 Community Based Initiative

Kuriyama has both potential and limitations like any other communities in Japan. Considering such conditions, the innovative community based initiatives are being implemented and introduced in this section.

3.1 'Kuriyama Tour'

The Kuriyama Tour was initiated by a working committee consisting of Local Vitalization Cooperators, Japanese Inn Association, shop owners, community based organizations, etc. in 2014. The tour was planned and executed by the committee. The program allows tourists to experience the rural outdoors such as trekking, canoeing and *satoyama* (mountain close to human settlement) experiences. Table 2 shows three types of tour programs offering 20 options with varying prices in different seasons. The tour guides were outsourced for trekking and canoeing because there are no professional tour guides in Kuriyama and it's necessary to keep the quality of services for the visitors. Some participants are the locals that have come to realize new attractions and have increased their interest in their local history and culture. Regarding the *satoyama* experience, the guide of *satoyama* experiencing tours are not outsourced and mainly consist of local community based organizations such as *Ishiyaki hozonkai* (Preservation society of Kuriyama stone barbecue), *Nikko kayabocchi no kai* (Preservation society of grassland), *Nikko shizen koubou* (Nikko natural workshop), etc. In this way, the tours increase the job opportunities and allow the locals to share their missions with the outside visitors. The committee consists of a variety of members from Kuriyama and the tour program is a good experience for them to work together closely sharing their knowledge, ideas and expertize in order to enhance the hospitality and satisfaction of the visitors.

Positive numerical results were seen soon after it started. The number of participants increased from 100 in 2014 to 127 in 2015 and 298 in 2016. In 2014, the accommodation came with the tour package at a special campaign cost of 2,000 yen and 100 people stayed in Kuriyama. Both in 2015 and 2016, approximately half of the tourists stayed in the accommodation in Kuriyama contributing to the local economy. Since 2015, more people applied than the available number; for instance, 376 people applied for the tour in total however 298 people are accepted due to the popularity of the tour in 2016. Therefore, the number of tours was 12 times at the first year of 2014 and it increased to 30 times in 2016 while the committee increased more options for the tourist (Shimotsuke Newspapers, 2013).

The other impressive numerical result is the repeat visitors. In 2015, 40 people out of 127 people are repeat visitors and 31% of people repeatedly joined the tour. In 2016, 133 participants out of 298 are repeat visitors meaning 45% of people repeatedly joined the tour. The increasing number of repeat visitors clearly shows high satisfaction of the tours and the increase in Kuriyama fans.

The other interesting feature is the promotion of the tour. After the tour ends, the participants' reports as well as their videos were posted on websites and Facebook, showcasing attractive features of Kuriyama. The number of posts increased nearly three times for three years from 2014 to 2016. Even with a few participants in the first year, the number of the participants increased in the second year due to posts on Social Network Service sites.

3.2 'Kamakura Matsuri'

The other initiative is called Kamakura Matsuri which is held annually between the end of January and early March in Yunishigawa. *Kamakura* is a traditional event from the northern part of Japan during *koshogatsu* (New Year's Day according to lunar calendar) in the middle of January in which people build snow huts for children to play in. There are five locations related to Kamakura Matsuri in 2018. At the main sites, over 20 *kamakura* are made and tourists can enjoy a barbecue inside. Outside of Kamakura, there are snow slides, food shops, *sake* tasting and concerts. The local people and volunteers start constructing the Kamakura from early January with 40

Table 2 Program of Kuriyama Tour

Tour programs	Month	Time/price (JPY)
Trekking		
1. To see the water fall at a height of 128 m	June	4 h./5,000
2. To see the fountain water and experience the hot spring with fountain water	June	4 h./5,000
3. At marshland at 2,000 m elevation to see the rare alpine flora	July and Sep	4 h./5,000
4. To the summit of the mountain to see Mt. Nantaisan and Kinu swamp	Oct	6 h./5,000
5. At Mt. Nikko and Kuriyama Dam to see autumn leaves	Oct	2.5 h./4,500
6. To the remains of mines, storehouse for gun powder, school, etc	Nov	4 h./5,000
7. In snow to see the fall with lunch and hot spring	Feb	3 h./4,500
8. In snow to see the frozen water fall with outdoor lunch	Feb	3 h./9,000
Canoeing		
9. At dam lake	May and Oct	3 times per day/5,000
10. With stand-up paddle board	Twice in July	Twice per day/5,000
Satoyama experience		
11. Bird watching and nature watch with lunch	May	Depends/1,500
12. Harvesting bracken and a walk	June	Depends/2,000
13. Making craft with deer leathers with lunch	July	Depends/3,000
14. Watching traditional lion dance and eat local food	Aug	Depends/2,000
15. Seeing the remains of gold mines in Nishizawa with lunch and hot spring	Sep	Depends/3,000
16. Experiencing traditional hot stone barbeque	Sep	Depends/5,000
17. Preservation of grassland with lunch	Oct	Depends/5,000
18. Experiencing hunters and traditional local food with lunch and dinner	Feb	Depends/12,000
19. Walking with snowshoes and harvesting maple syrup	Feb. and Mar	Depends/3,000
20. Watching stars at summer night at Kawamata Dam	Aug	Depends/500

tons of snow in Kuriyama. In another location, there is a night illumination of a thousand small *kamakura* that are 40 cm in height along the river, the candles are lit up inside of *kamakura*, creating fantastic night view which was designated as a Japan heritage. The other attraction of Kamakura Matsuri is the outdoor experience at snow parks; riding on a sleigh and going down the snow slide for small children and snow rafting for adults (see Fig. 2).

Kamakura Matsuri started 25 years ago. When the event started, Kamakura Matsuri was small-scaled, but recently more and more people visit mainly due to the designation as a Japan heritage. The number of tourists in Kuriyama was less during the winter season in the past; in fact, until 30 years ago, the months between January to March were the least busy months of local tourism throughout the year due to the excessively deep snow and difficult accessibility. However, Kamakura Matsuri became one of the most attractive events in Kuriyama recently. In fact, the number of visitors to Kuriyama in February was 15,759 people 35 years ago and

Fig. 2 Pamphlet of Kamakura Matsuri

now increased to 32,816 people in 2016. February is the fourth busiest month after August during summer holiday and October and November for colored leaves season (Nikko City, 2016). Therefore, Kamakura Matsuri has become a special feature of winter in Kuriyama.

3.3 '5 Kuri'

5 kuri (an introduction of Kuriyama in five second) is the only website offering tourist information of Kuriyama. It was developed and managed by a Local Vitalization Cooperator who came to live in Kuriyama in 2012. Although it normally takes much more than five second to find and read the information, the website is very informative covering tourist attractions of all areas, events, culture and history, gourmets, Japanese Inn, and the access (Aoyama, 2017). Nikko Tourism Association offers the information of Nikko City including Kuriyama, but 5 *kuri* is the only information source covering only Kuriyama.

3.4 Possible Target: Foreign Tourists

Another interesting initiative is the development of potential foreign tourists. This initiative was started two years ago by a Local Vitalization Cooperator. Foreign tourists were target as there is less competition in the foreign tourism market and hence more possibility to grow compared to the Japanese tourist market. In 2016, only 1,293 foreign tourists stayed at an accommodation in Kuriyama accounting for only 0.5 percent out of 242,863 entire tourists who stayed in Kuriyama. Nikko area had 54,916 foreign tourists in 2016, accounting for over 40 times more than Kuriyama (Nikko City, 2016). In this sense, there is more potential in attracting foreign tourists to Kuriyama. The traditional culture and history, landscape and rural lifestyle of Kuriyama can attract foreigners based on this experience. Also, the famous tourist locations in Tokyo and Kyoto are recently reaching saturation because more and more foreign tourists come to Japan, therefore there is a trend of diversifying tourist locations.

A Local Vitalization Cooperator organized a trial package tour for 21 foreign tourists to promote foreigners to visit and stay in Kuriyama in 2016 and 2017, received eight Jordanian trainees in community tourism in 2016 and 22 foreign students of Utsunomiya University to discuss community development and tourism. It is too early to see concrete results, however was able to gain an insight into the local peoples' capacity and services as well as points to be improved to receive foreign tourists.

4 Lessons Learned and the Way Forward

4.1 Utilization of Local Resources

One of the typical but important lessons learned from the case of Kuriyama is the utilization of local resources. There is a variety of local resources in Kuriyama. One is natural resources including landscapes, hot springs, climate, mountains, trees, snow, etc. enabling visitors to enjoy throughout the year. Not only tangible resources, but intangible resources such as culture and tradition are important resources for the local community to promote eco-tourism. Such resources including rich culture and long history, festival, traditional lion dance, etc. are handed down from ancestors for a long time. Human resources are important assets of the community. Although the number of local people are decreasing every year, there are still human resources who have knowledge of tradition and culture rooted in the community. Such both tangible and intangible resources are not easily replicable and also not replaceable in short time once they lose them.

Okoso proposes necessary approaches to vitalize a community which are:

(1) uncovering community resources,
(2) understanding the value of community resources by the local people, and
(3) selling the commodities and services of local resources (2016, 97–98).

The local resources found in Kuriyama are not something new and have existed in the community for long time. As a result of that; however, they are not often considered as resources, especially by local people. In this sense, the rediscovery, re-appreciation and the utilization of the local resources are the important processes for the vitalization of the community towards sustainable tourism.

4.2 Effective Marketing Strategies

Although the resources have existed for a long time in the community, the approach to the utilization of resources must be modified. The Local Vitalization Cooperators coordinate and offer ample options for tourists to enjoy making a choice in the initiatives of Kuriyama Tour. Each program offers the opportunities to actually experience local resources. The experience provided by professional guides or local people are real and not artificial and the tourists feel close to the local community.

The promotion of tourism has recently changed rapidly. Social Network Service such as websites and Facebook have a strong influence in attracting customers, especially for people who have never visited Kuriyama. 5 *kuri*, the only website offering tourist information of Kuriyama, contributes to increase the fans of Kuriyama. Not only for the possible tourists or short-term fans, but also repeaters, the people repeatedly check the website and make their comments through Facebook and come back to Kuriyama to experience in reality.

4.3 Partnerships Within the Community

This case shows the cooperation of local people and outsiders. The local people include a variety of people such as hotel owners, shop owners, members of Japanese Inn Association, community based organizations, local administrators of tourism, etc. The local people themselves have a limitation because over 40% of them are above 65 years old. Fewer people are entering the labor force in Kuriyama which makes it difficult to start new programs or new methods of promotion and conduct events and festival every season and even to maintain the hotels and shops. Can the local people construct *kamakura* with 40 tons of snow on their own? Can they manage with the influence of Social Network Service? People from outside can play an important role in coordinating, managing, and supporting the local efforts. Outsiders include Local Vitalization Cooperators, families and relatives who live outside of Kuriyama, volunteers, academics and tourists. It is difficult to promote tourism by one stakeholder. For instance, if the hotel is great, but there are no attractions or shops or effective work force, the hotel might gradually degrade. As Kuriyama tour became a good experience for the different stakeholders in the committee to work together closely, the partnerships within the community are necessary in order to enhance the hospitality and satisfaction of the visitors.

4.4 Paradigm Shift from Limitation to Potential

The potential and limitation are explained in the background of the case. The challenges which Kuriyama faces include depopulation and decreasing labor force, poor accessibility and threats of nature such as serious winters and natural disasters which are commonly observed in many marginalized rural communities in Japan. To break this situation, the people who are involved with tourism and community development, made a paradigm shift of thinking. In spite of fewer tourists in winter before due to the serious snowfalls and less accessibility to Kuriyama, Kamakura Matsuri became one of the most attractive events throughout the year.

Lack of accessibility was raised as one of the limitations of Kuriyama, according to a Local Vitalization Cooperator, however, it is an advantage that adds to the value of unexplored regions. Tourism has become diversified as some tourists want to experience mysterious land. Kuriyama is called the last resort because it is located at the most north in Kanto regions which include big cities such as Tokyo, Saitama and Kanagawa. Therefore, limitation of less accessibility is shifted its paradigm to a potential.

4.5 Concluding Remarks

The notion of sustainable tourism has been widely recognized since the 1990s. This chapter introduces the case of Kuriyama, Nikko City in Japan, which has been developed in rich culture and history. While Kuriyama has magnificent landscapes every season, challenges include the lessening of the labor force, natural disasters, poor accessibility and the merge into Nikko City that hinder local tourism to become sustainable. Several initiatives were taken by the local people and outsiders. The lessons learned from the initiatives present both tangible and intangible utilization of local resources, effective marketing strategies, the partnership within the community and the paradigm shift from limitation to potential. The raised summary of the lessons learned presents a guide for a better sustainable tourism. The community based initiative introduced in this chapter is just one of the encouraging initiatives out of repeated trial and errors which stakeholders have made. It is hoped that the lessons that emerge will be relevant to anyone involved in tourism as well as community development in Japan and in the world.

Acknowledgements This paper represents the timeless efforts of all the stakeholders who have promoted tourism in Kuriyama for the purpose of sustainable communities. The author wishes to thank Mr. Hikino Goichi and Mr. Aoyama Hiroyuki, Local Vitalization Cooperators in Nikko City and Ms. Ban Hiromi, Landlady of Yunisikawa Inn for sharing the information and appreciate their commitment to enhance tourism in Kuriyama.

References

Aoyama H. (2017). *5 kuri*. Retrieved Jan 26, 2018, from https://9r8m.jp/. (in Japanese).
Butler R. (1999). Sustainable tourism. *Tourism Geographies*, 1(1).
Cabinet Office, Japanese Government. (2017). Aging situation. *Aging White Paper*. (in Japanese).
Nikko City. 2010, 2011, 2012, 2013, 2014, 2015, and 2016. Statistic book. (in Japanese)
Okoso M. (2016). *Community development and tourism through local platform*. Kyoto: Gakugei Shuppansha (in Japanese).
Shimakawa T. (2009). Important issues for tourism. Tokyo: Doyukan (in Japanese).
Shimotsuke Newspapers 2013. Closing down Japanese Inn: Giving up the recovery. Shimotsuke Newspapers. August 1, 2013 (in Japanese).

Eco-Tourism Project in Hong Ha Commune: A Case Study of Sustainable Community Initiatives in Vietnam

Tung Ngoc Nguyen, Miki Yoshizumi, Hirohide Kobayashi, and Van Anh Ngoc Le

Abstract This chapter presents a case study of community-based ecotourism development in Hong Ha commune, a remote mountainous area in Thua Thien Hue province, Vietnam. Drawing on field surveys, interviews with local residents, government officials, and tourists, the study explores how ecotourism initiatives have contributed to local livelihoods, cultural preservation, and sustainable development. The chapter highlights the self-organised structure of the ecotourism program, the role of traditional community houses as social and cultural hubs, and the involvement of ethnic minorities in tourism activities. Despite its successes, the project faces several challenges, including limited infrastructure, environmental concerns, and a need for improved strategic planning. The findings offer insights into the potentials and limitations of grassroots ecotourism in rural Southeast Asia and provide recommendations for future improvement and policy support.

Keywords Community-based tourism · Rural development · Eco-Tourism · Ethnic minority communities · Local governance

---

T. N. Nguyen (✉) · V. A. N. Le
Faculty of Architecture, University of Sciences, Hue University, Hue, Vietnam
e-mail: Kts.nguyentung@hueuni.edu.vn

V. A. N. Le
e-mail: Lnvanh@hueuni.edu.vn

M. Yoshizumi
College of Gastronomy Management, Ritsumeikan University, Kyoto, Japan
e-mail: miki-yo@fc.ritsumei.ac.jp

H. Kobayashi
Graduate School of Global Environmental Studies, Kyoto University, Kyoto, Japan
e-mail: kobahiro@archi.kyoto-u.ac.jp

M. Yoshizumi et al. (eds.), *Navigating Local Sustainability in Food, Community, and Innovation*, https://doi.org/10.1007/978-981-96-8963-7_8

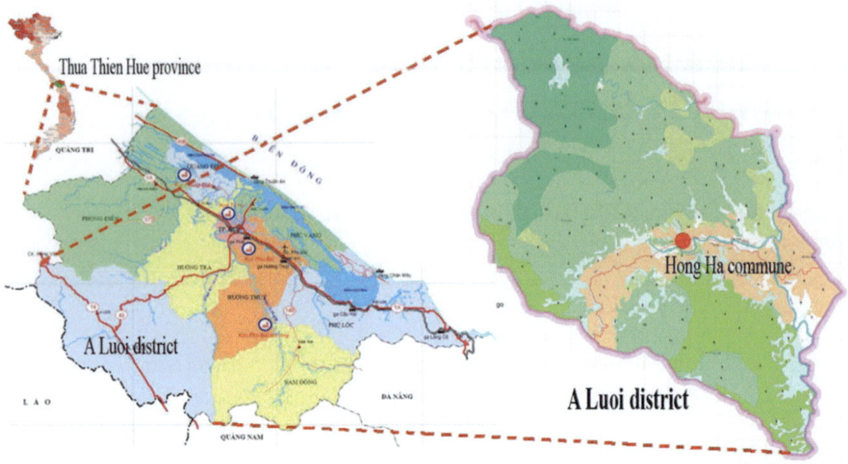

Fig. 1 Map of Hong Ha

1 Scientific Background

Hong Ha commune is located in the mountainous area of A Luoi district in Thua Thien Hue province, Vietnam (Fig. 1).[1] The commune has five hamlets and some ethnic minorities such as Katu, Ta Oi, and Pa Co (Tran, 1984; Luu, 2007; Dang et al., 1993). It has a wide range of ecotourism resources, such as forests, traditional community houses, Parle waterfall, A Zoi stone, traditional dance forms and local cuisine. The commune implemented an ecotourism project in 2016, with the people's committee and locals selecting some sites for the project, such as Parle waterfall, traditional community house and A Zoi stone (Figs. 2, 3 and 4).

Ecotourism in Hong Ha has developed expeditiously based on the contributions of the local community and government in the commune. In 2016, many tourists visited Hong Ha, and the commune plans to develop additional ecotourism sites and homestay programmes. This research explores the potential and initiatives of community-based ecotourism in Hong Ha as a successful case study for sustainability in Thua Thien Hue province in particular and Vietnam in general.

[1] According to the Resolution No. 175/2024/QH15 dated November 30, 2024 on the establishment of Hue city under the Central Government of the Chairman of the National Assembly, Hue was a Centrally Governed City with the entire area and population of Thua Thien Hue province. This Resolution takes effect from January 1, 2025. This article refers to place names and administrative boundaries based on the status prior to 2025

Fig. 2 Parle waterfall in Hong Ha

2 Objectives and Methods

This study has three objectives:

- Evaluate the ecotourism situation in general in Thua Thien Hue province, Vietnam
- Study the potential and initiatives for developing community-based ecotourism in Hong Ha commune
- Provide preliminary recommendations to further improve ecotourism.

Two types of documents are used to examine the objectives:

Secondary documents are obtained to understand the background, situation and previous ecotourism projects in Hue city. These include legal documents, reports, newspapers, magazines, basic maps and other relevant data.

Primary documents are derived from a case study field survey of the Hong Ha commune. Surveys were conducted in Hong Ha between 2016 and 2017. Interviews were conducted with the leader of the Hong Ha commune government, staffs of the ecotourism association, tourists and residents of the commune.

Fig. 3 Community houses in Hong Ha

3 Situation and Evaluation of Ecotourism in Thua Thien Hue Province

Thua Thien-Hue province is located in Central Vietnam, and comprises four zones: mountains, hills, plains and lagoons, separated from the sea by sandbanks. With 128 km of beachline and mountains covering more than half of the total surface area, the province lies along the western and southwestern borders.

The wealth and diversity of cultural and natural resources of Thua Thien Hue are considered important prerequisites for the development of various types of tourism, such as cultural and ecotourism, marine, medical and leisure tourism (Table 1). The tourism resources of Thua Thien Hue are concentrated in three regions: Hue city and neighbouring regions (area of Hue city and Thuan An beach), the southern region (Canh Duong beach, Bach Ma mountain, Lang Co beach, Hai Van mountain pass and Son Tra peninsula) and area of A Luoi district and Ho Chi Minh trail (Tran, 2012). Each area offers various tourism packages with its own typical nuances. Additionally, with its varied terrains carrying specific characteristics of geological structure, Thua Thien Hue has abundant natural resources to facilitate the development of various types of activities associated with nature. Its natural resources include sea, lagoons, national parks, rivers, lakes, hills, mountains, waterfalls and hot springs.

Fig. 4 A Zoi stone in Hong Ha

Table 1 Number of tourists who visited Hue in 2012–2016 *source* Conservation and Center 2007

Year	2012	2013	2014	2015	2016
Foreign visitors	861,239	947,514	931,894	1,075,543	1,198,350
Domestic visitors	911,299	924,989	1,000,919	971,418	974,487

3.1 *Ecotourism Development Status in Thua Thien Hue Province*

In recent years, ecotourism has developed in Thua Thien Hue, especially in the Tam Giang Lagoon area and mountainous districts such as A Luoi and Nam Dong. Natural landscapes and unique local culture have increasingly attracted tourists and opened new livelihood development opportunities for local people.

However, research on fundamental ecotourism resources of Thua Thien Hue has been limited, unsystematic and incomprehensive. Although some preliminary assessments have been conducted (e.g., WWF, 2002), no official studies have been carried out so far to develop ecotourism in the province. Regional planning mission can offer potential for ecotourism development in Thua Thien Hue, but it is still nascent.

3.2 *The Status of Exploitation of Natural Tourism Resources*

Although natural tourism resources in Thua Thien Hue are valuable and have a great appeal to domestic and international tourists, the exploitation of these resources to serve tourists has been inadequate and unstable, mainly focusing on cultural tourism in recent years (Ton, 2014).

Natural tourism resources, including Canh Duong, Lang Co and Thuan An beaches, Bach Ma National Park, Thanh Tan and My An hot springs, Suoi Voi, Nhi Ho lake and Suoi Tien stream, have been attracting an increasing number of visitors, primarily domestic visitors, during weekends. New ecotourism packages such as Bach Ma National Park, My An and Thanh Tan hot springs and the urban vegetable garden tour 'Explore the Ngu Ha River' have been received positively. This has contributed to the diversification and enrichment of various tourism products, gradually forming new tourism destinations, reducing the pressure in the heart of Hue city and prolonging the length of guests' stay.

In addition, the Tam Giang-Cau Hai lagoon is considered one of the largest in the Southeast with its diverse ecosystem, but it only takes advantage of the water surface area for aquaculture. Meanwhile, it has the potential to be developed further considering that it has many resources that can be utilised for different types of ecotourism, such as boating, fishing, adventure sports, etc. However, this potential has been overlooked and lacks adequate investment. Among the areas exploited for tourism purposes, only two regions serve ecotourism development: Bach Ma National Park and the urban vegetable garden in Hue. Bach Ma National Park has drawn appreciation for its quality and investment planning of ecotourism development. The 'vegetable garden' in Hue received widespread appreciation during the Hue Festival 2000; however, it has not been maintained well thereafter.

3.3 Regional Planning for Ecotourism Development

In terms of space, three development areas with different ecological terrains can be considered, as follows:

Coastal and lagoon areas: Ecosystem of Tam Giang-Cau Hai, belonging to O Lau river estuary, is characterised by a high density of migratory water birds in the spring, eco-cultural villages along the coast and a lagoon area.

Eco-regions of Bach Ma-Nam Dong-A Luoi: Bach Ma National Park is located about 40 km from Hue to the south and has a cool climate and fresh air, since it is close to the sea. It also has several rare species of animals and plants with scientific significance and an ideal terrain for sporting events such as mountain climbing and expeditions.

On this route, there are two eco-cultural-historical zones: A Luoi and Nam Dong, which have beautiful landscapes that are rich in nature and cultural resources. Their tropical forests have enormous socioeconomic significance for both the province and local communities. These areas are also home to ethnic minority groups, including Co Tu, Pako and Ta oi, which have special traditional cultures.

Hue city and surrounding areas: There is a potential to develop ecotourism owing to traditional gardens, vegetable gardens surrounding the citadel and ancient villages.

4 Ecotourism Programme in Hong Ha Commune and Its Initiatives for Development

4.1 Progress of Ecotourism in Hong Ha Commune

Hong Ha is located upstream of the Bo river, one of the three main rivers flowing into Tam Giang lagoon. The commune is 40 km from Hue city on the main road to A Luoi district. It has an area of 14,100 ha and its population was 1,766 in 2016. Ethnic minorities have settled here and follow various traditional cultivation styles. Hong Ha is one of the 2,000 poorest communes in the country and among the 16 poor communes in A Luoi district.

The local government of Hong Ha commune initiated ecotourism in March 2016 as a new income source for local people (People's Committee, 2016a). Initially, Parle waterfall, located 1 km from the People's Committee of Hong Ha commune on National Road 49, was the main local resource for ecotourism programmes (Fig. 5). The commune constructed parking spaces, washrooms, introduction boards, more than ten stilt houses and spaces for swimming and relaxation at the venue. A café was created in a traditional community house located next to the office of Hong Ha People's Committee.

Fig. 5 Advertisement of ecotourism in Parle waterfall

To facilitate ecotourism, an organisation consisting of 20 local people was established (People's Committee, 2017) (Fig. 6). Generally, three members of the Hong Ha People's Committee (chairman, vice chairman and secretary of the Communist Party) are responsible for managing all ecotourism activities. Two sections were set up to control various issues: personnel, cultural, finance, planning and accounting. Other members worked in different fields in the activity group, such as services, security, kitchens and tickets, among other duties.

The initial investment was provided by the People's Committee of Hong Ha commune. At least 30% of total income from ecotourism is distributed to the People's Committee as management fee and for advertising. About 70% of the income is used for the sale of food products, providing allowances to members and a contribution fund.

The traditional community house was originally used for community activities, such as a classroom for learning about wood protection, library, community meeting place and space for traditional dance performances and ceremonies (Kobayashi et al., 2008). However, these activities are rare. The commune initiated the establishment of a WiFi café and a small library in a traditional community house (Fig. 7). According to Mr. Ho Viet Luong, chairman of the Hong Ha People's Committee, the café in the community house was an initiative to expand the scope of activities there. A staff member of the commune works in the traditional community house to sell coffee and other drinks, and the staff can check the condition and clean the house every day.

Fig. 6 Structure of ecotourism project members in Hong Ha commune

This activity can help conserve the house, while the staff can earn additional income from the sale of drinks. Meanwhile, local people and tourists can stop briefly during the trip and relax.

The commune has also created other ecotourism sites based on tourists' needs. If tourists wish to experience the mountain, natural landscape, lifestyle of local people and A Zoi stone, a programme can be arranged accordingly. To promote ecotourism,

Fig. 7 Two book cases in a community house

the Hong Ha commune government signed a contract with destination management company HGH that has advertised Hong Ha's ecotourism in Hue city. Accordingly, the company collaborated with Hong Ha commune to organise activities to popularise ecotourism in Hong Ha. For example, HGH published Hong Ha ecotourism details in the newspaper and through television broadcast. Furthermore, the company organised an event in Hong Ha in 2016 where local people wore traditional attire and rode bicycles around Hue city to introduce the ecotourism programme. A social network page (Facebook) was set up on Parle waterfalls. The web page has received 1,755 likes and has been accessed by 3,615 people so far.

The community house was found to be disintegrating around August 2016, with a leaking roof and some wooden components being destroyed by termites. Simultaneously, the need for providing accommodation for tourists in Hong Ha continued to increase. Therefore, in March 2017, the commune repaired the structure and introduced more facilities such as bedding on the upper floor and a café and restaurant on the ground floor (Figs. 8 and 9). New washrooms were constructed in the rear portion of the building. The construction was supported by the Hong Ha People's Committee and labourers from five hamlets. The homestay opened in August 2017. In this month (August), 30 tourist groups visited and stayed at night in the community house. The total income from the homestay was VND 11,940,000 (approximately USD 527. According to the webpage https://www.vietcombank.com.vn/ExchangeRates/ on 25 January 2018, USD 1 equals VND 22,675).

Fig. 8 The café was set up on the ground floor of a community house

Fig. 9 Homestay in a community house

In addition, some portions of the stilt houses and infrastructure system in Parle waterfall were destroyed by a typhoon that struck in January 2017. The commune planned to reconstruct some stilt houses, roads and toilets in Parle waterfall, mainly based on community labour. Currently, there are 22 stilt houses (the commune repaired seven and constructed two, while five new stilt houses were built by local people. The others were constructed in 2016), two advertisement boards and five small bridges at Parle waterfall.

A smart initiative for sustainable ecotourism is the commune's long-term plan to develop ecotourism programmes. The commune expressed concerns about the difficulty of collecting wood materials to replace some components of community houses in the future. The People's Committee encouraged local people to plant Sao Den trees (Golden Oak) in their gardens. Although the quality of Sao Den is not as good as that of Kien Kien (Peck tree), it grows straight and is easy to maintain compared to Kien Kien. Additionally, 10 households have planted Cọ tree (palm tree) that can be used later when replacing the roof of the community house (each household plants 20 Cọ tree). Moreover, the commune collaborated with the Hue Association for Friendship to organise training courses for members of the ecotourism organisation. They also encouraged the villagers and patriarchs to perform traditional weaving and dance for tourists in the community house. In primary schools, patriarchs can perform and also teach traditional dance forms to students. The values of traditional dance can be passed on from generation to generation.

4.2 *Interview Surveys*

To understand the potential for ecotourism in Hong Ha commune, interviews were conducted with government staff, local people and tourists. The following section presents the data analysis and survey results.

4.2.1 Interviews with Local People

Seven local people living in Hong Ha commune were interviewed to obtain information about ecotourism (Fig. 10). They were aged between 30 and 60, and four were female.

Five respondents were participating in an ecotourism project in Hong Ha. Among them, four were part of the cuisine team, while one worked as a service staff at Parle waterfall. The average income generated by the ecotourism project was VND 2,900,000 (approximately USD 128). Some of the reasons cited by participants were 'want to develop Hong Ha' and 'earn more income'. Two people were not involved in the project because they lacked sufficient resources to join, or their families did not want them to.

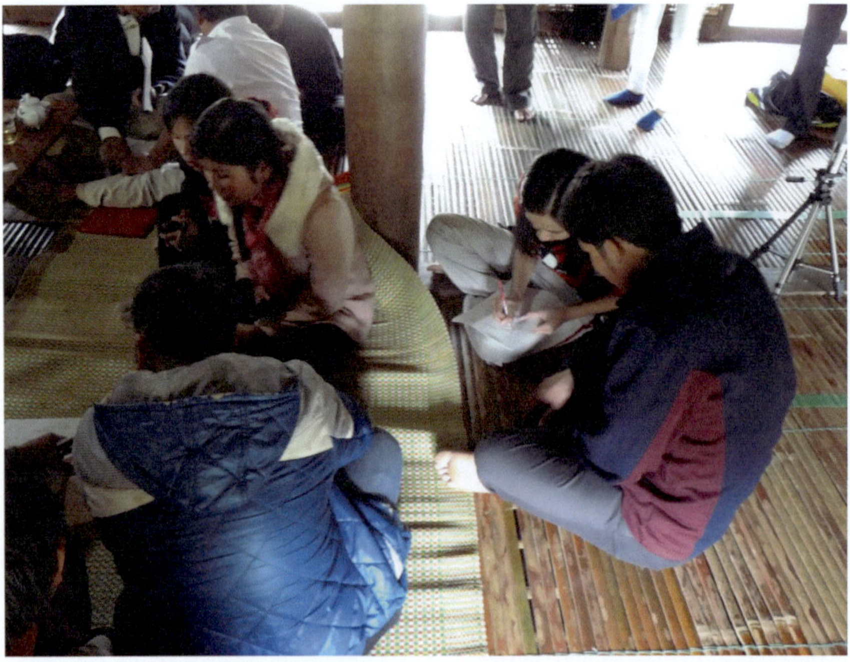

Fig. 10 Interviews with local people in a community house

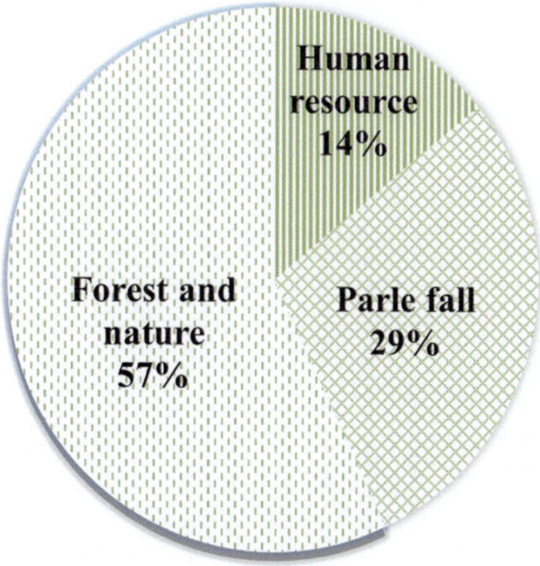

Fig. 11 Hong Ha's resources for ecotourism

To develop ecotourism in Hong Ha, four people chose forest and nature (Fig. 11). All of them claimed that ecotourism in Hong Ha should be developed expeditiously because it can create jobs for local people, as well as provide various natural resources.

Figure 12 illustrates respondents' opinions on various issues regarding ecotourism in Hong Ha. All the respondents either agreed strongly or moderately that Hong Ha has immense potential for developing ecotourism, but infrastructure should be improved (traffic and road conditions). They also agreed that the traditional community house is an important element for ecotourism in Hong Ha, and that more activities should be organised there, while the landscape surrounding the house should be developed. The respondents also acknowledged the current function of the traditional community house, which also serves as a café, to hold traditional dance performances and other community activities.

Most respondents stated that guesthouses and homestays should be constructed as part of ecotourism in Hong Ha. Guesthouses should be stilt houses or similar to traditional community houses. Additionally, some respondents believed that the food at Parle waterfall was expensive, and also recommended the construction of a safe pathway for children. They also suggested that a pathway be built to connect the waterfall and traditional community houses, so that visitors could go for a swim, relax and then return to the community house for food and coffee.

A Zoi stone is another ecotourism site in Hong Ha. There is a legend about A Zoi stone (more details about the legend are given in Nguyen (2016)). Most respondents strongly or moderately believe in the legend of the A Zoi stone. They were also

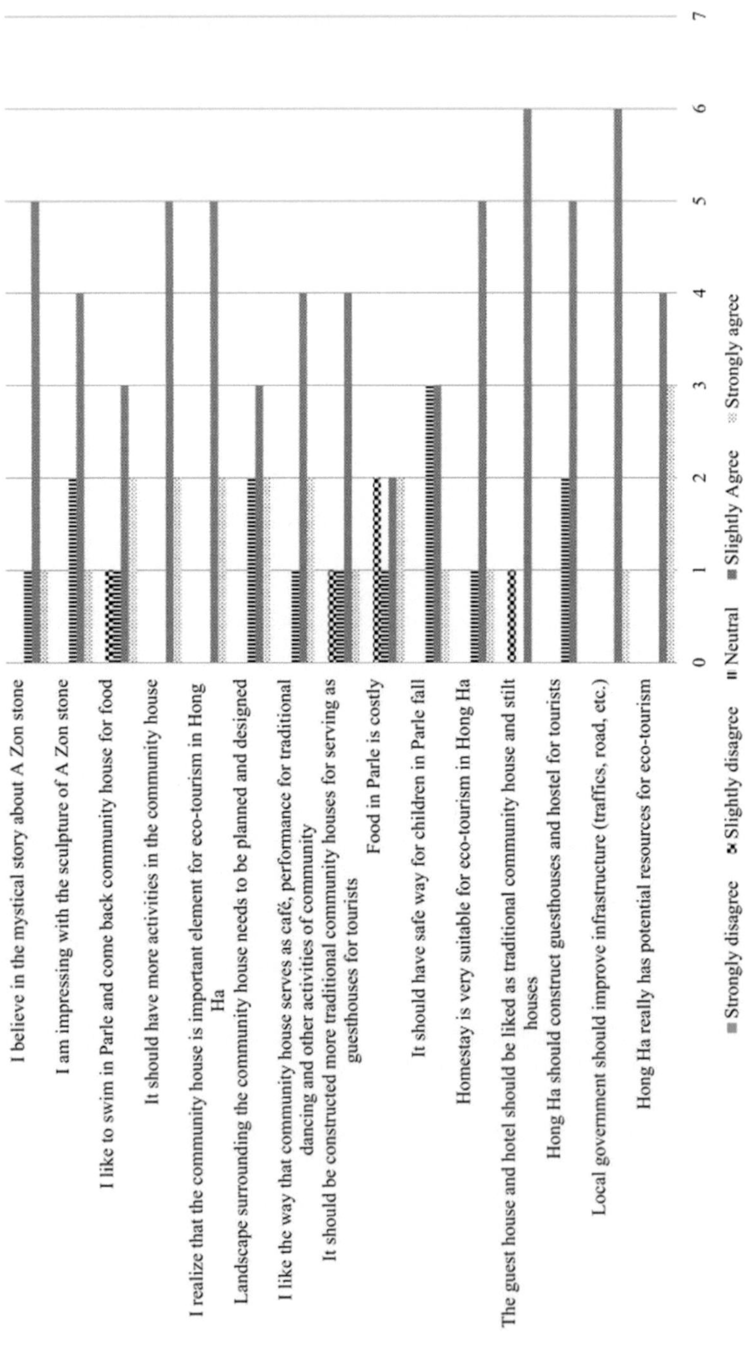

Fig. 12 Local residents' assessments of ecotourism challenges and opportunities

impressed by the stone sculpture. They believe that the natural landscape, Parle waterfall and A Zoi are the sites that attract tourists to Hong Ha.

With regard to improving ecotourism, the respondents suggested that Hong Ha seek support and investment from government and organisations to expand tourism projects. Improving the infrastructure and offering training courses for tourism services are among the other suggestions.

4.2.2 Interviews with Local Government Staff

Twelve government staff members were interviewed to garner their opinions on the ecotourism project in Hong Ha. Some of them held important positions in local government, such as Mr. Luong (chairman of the Hong Ha People's Committee), Mr. Nguyen Hoai Nam (chairman of the Association of the Elderly) and Mr. Le Van Hoi (vice chairman of the Hong Ha People's Committee). They provided useful information and future plans for improving ecotourism.

Most of them mentioned that the People's Committee makes suggestions regarding ecotourism projects in Hong Ha, except for two respondents who mentioned that Mr. Luong was responsible for making decisions. All respondents participated in the ecotourism project and assumed different positions, such as management, consultant, advertising and service. They participated in the project because they wanted Hong Ha to develop (Fig. 13). Other reasons include higher income, work satisfaction and responsibility. Regarding the potential resources for ecotourism development in Hong Ha, most respondents claimed that Parle waterfall, community house and A Zoi stone were the main resources (Fig. 13). Others include human resources, traditional food and products. The respondents attributed the expeditious development of ecotourism to the local government positively creating conditions for ecotourism, local people earning more income and the various natural resources in Hong Ha (Figs. 14 and 15).

Figure 16 illustrates respondents' opinions regarding various issues related to ecotourism in Hong Ha. Ten local government staff members strongly or moderately agreed on some issues, such as Hong Ha having potential resources for improving ecotourism and the need to include additional activities, enhance promotions and construct guesthouses for the project; introduce training courses for project staff to improve their tourism skills; and the need for local governments to improve infrastructure (transport facilities and better roads to tourist sites, etc.).

The respondents also agreed that the community house plays an important role in the ecotourism project in Hong Ha, and that it should include more activities, while the surrounding landscape should be developed. Some government staff members did not agree on the construction of guesthouses, such as traditional community houses or stilt houses, because they believed that it would be difficult to preserve. The respondents also highlighted the lack of washrooms at Parle waterfall and complained that the food was expensive.

According to six respondents, natural resources and landscape were the main factors that attracted tourists to Hong Ha, while other factors were Parle waterfall

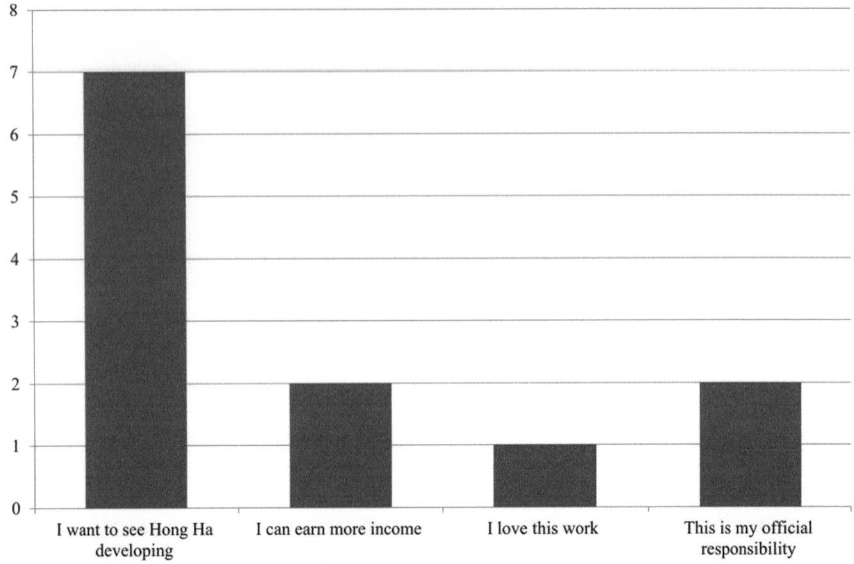

Fig. 13 Reasons for joining ecotourism projects

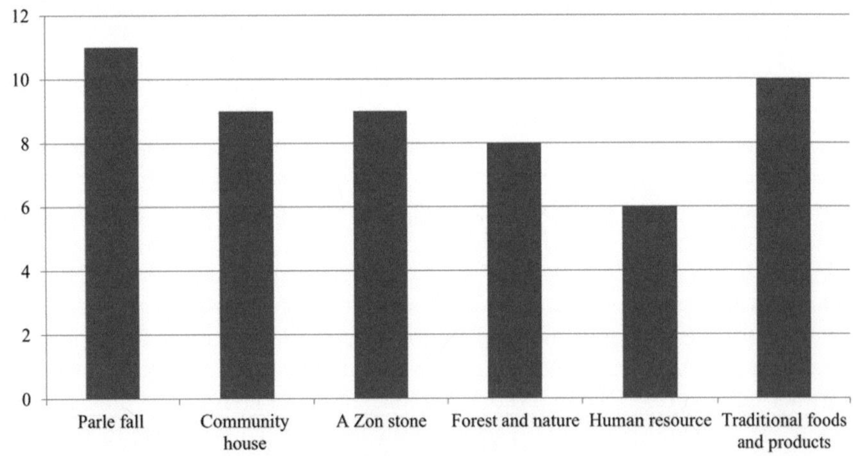

Fig. 14 Resources for ecotourism in Hong Ha

and A Zoi stone. The respondents also mentioned some potential sites that could be used for ecotourism, such as the Pilo stream, A Rum stream and natural forest. They believed that Hong Ha should introduce strategies to popularise tourism. They also agreed to join a training course on tourism and collaborate with Hue University for the future planning of ecotourism in Hong Ha.

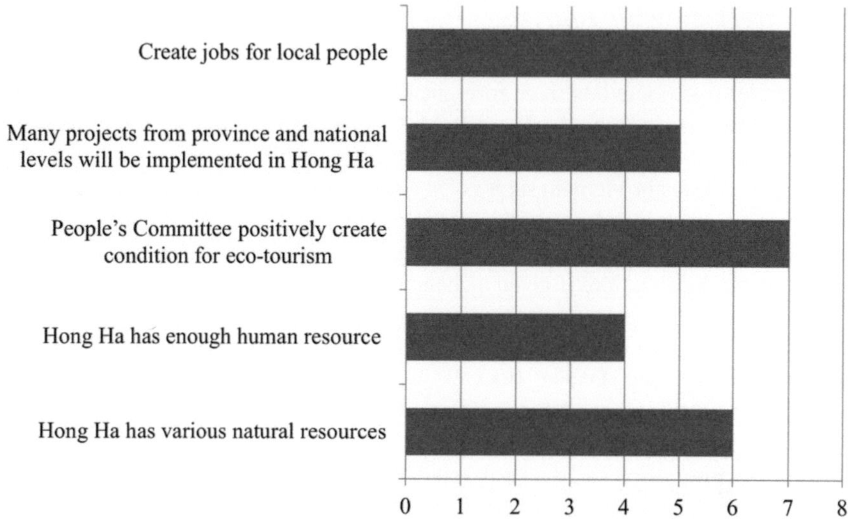

Fig. 15 Reasons for the expeditious development of ecotourism in Hong Ha

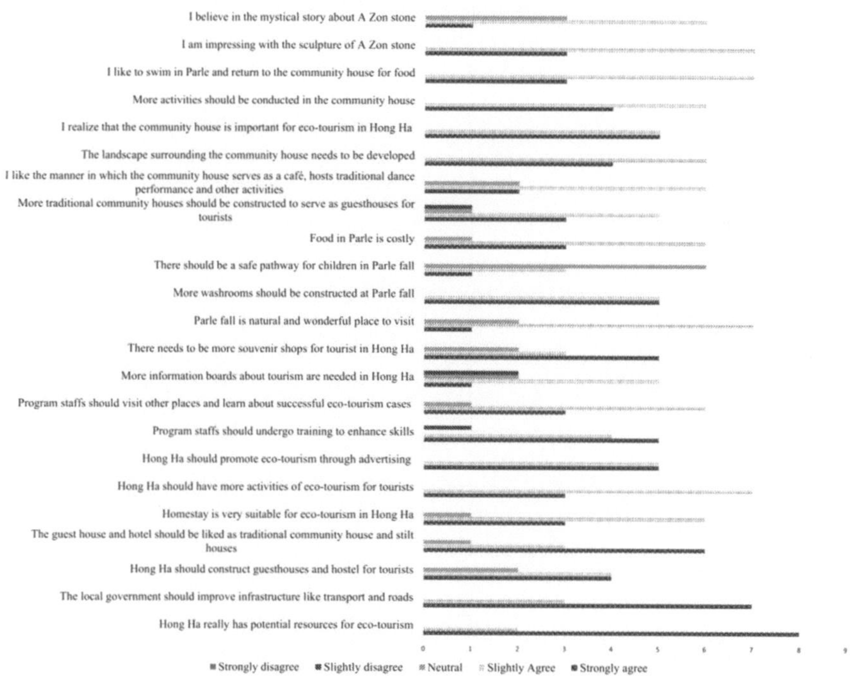

Fig. 16 Local government staff's views on ecotourism development strategies

4.3 Interview with Tourists

Fifty-one tourists, including three foreigners, were interviewed to garner their opinions on ecotourism in Hong Ha (Fig. 17). Most of them spent a day in Hong Ha except for two who stayed for two days. These foreigners spent two days in Hue and later visited other places in Vietnam such as Da Nang, Hoi An, Ha Noi and Ho Chi Minh. Eleven and eight respondents said they mainly visited Hong Ha for Parle waterfall and the community house, respectively (Fig. 18). Some also expressed interest in exploiting the local culture if given a chance to visit residential houses.

According to the survey, most tourists wanted to enjoy traditional local food (13 people), swim in Parle waterfall (nine), drink at the café and rest in the community house (seven) in Hong Ha (Fig. 19).

Nine respondents claimed that they decided to visit Hong Ha based on recommendations from their friends (Fig. 20). Others visited Hong Ha based on information from the Internet, such as Facebook (social network) and coincidentally. Cultural experiences and natural and ecological sites were motivating factors that convinced them to visit Hue (Fig. 21). Some respondents did not care much about the heritage sites in Hue.

Figure 22 shows tourists' opinions about ecotourism in Hong Ha. Most (nine people) strongly agreed that Hong Ha should have a homestay. Seven and five people

Fig. 17 Interview with tourists

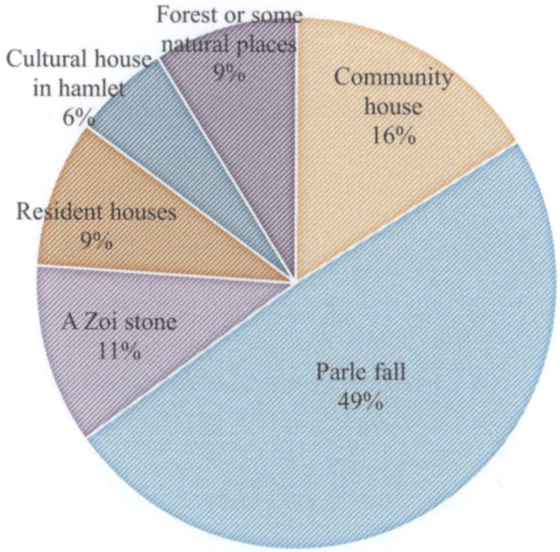

Fig. 18 Sites to visit in Hong Ha

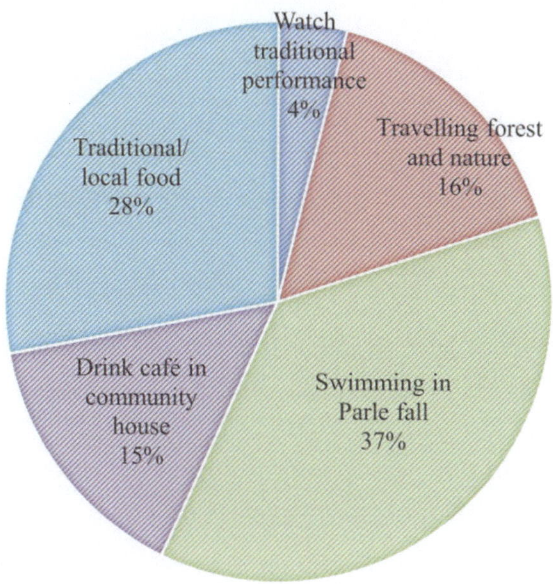

Fig. 19 Respondents' activities in Hong Ha

strongly and moderately agreed, respectively, that Hong Ha should enhance advertising strategies regarding ecotourism. Most either strongly or moderately agreed that Hong Ha has potential resources for developing ecotourism, that guesthouses should be constructed and infrastructure should be improved. Some tourists believe that no more souvenir shops should be constructed for tourists in Hong Ha.

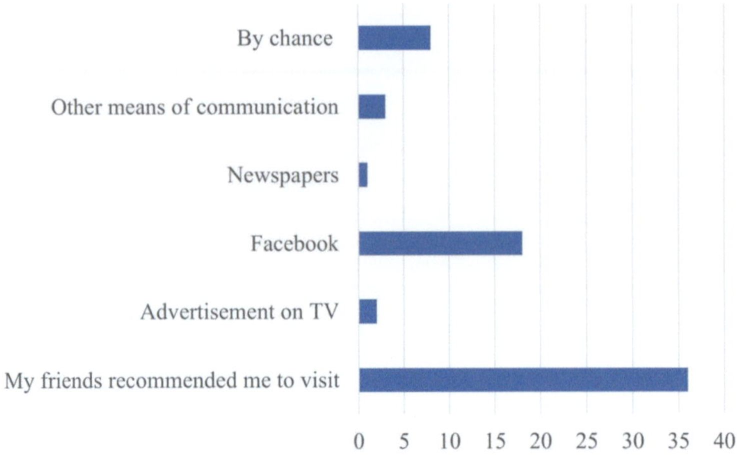

Fig. 20 Various means by which tourists learnt about Hong Ha

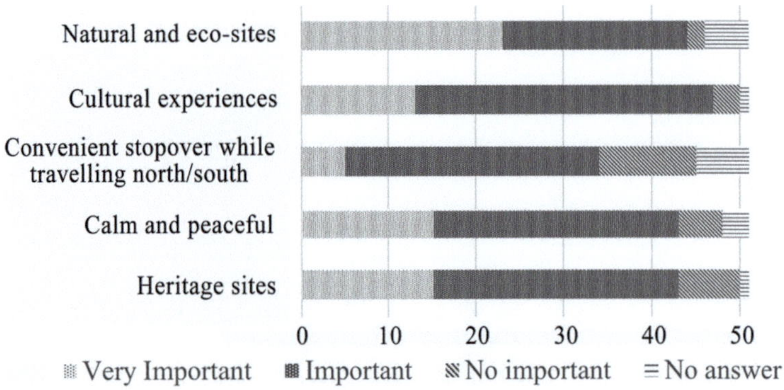

Fig. 21 Motivating factors for visiting Hong Ha

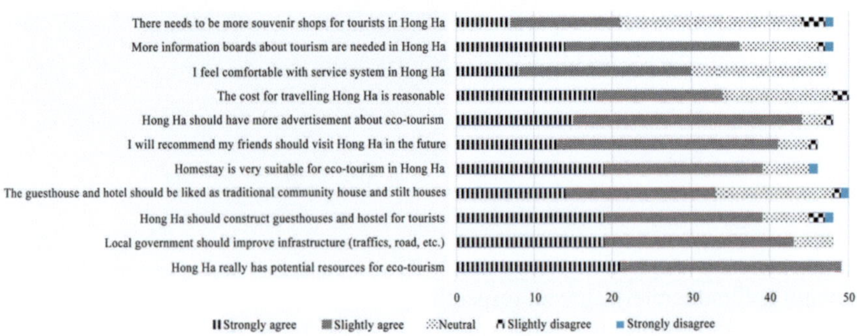

Fig. 22 Opinions of tourists regarding ecotourism in Hong Ha

All tourists strongly or moderately agreed that Parle waterfall is a natural and beautiful place to visit. They also mentioned that there were few washrooms in Parle waterfall. The tourists also believed that the architecture of the community house in Hong Ha was precious but more activities should be introduced. Regarding A Zoi stone, four tourists were impressed with the sculpture, but only one believed in the legend of the A Zoi stone.

Most of the tourists were satisfied with Hong Ha's service level. Various elements of Hong Ha attract tourists, such as the Parle waterfall, community houses, traditional food and natural landscape. They also offered advice for improving ecotourism in Hong Ha such as infrastructure improvement, training courses for staff and concrete organisations for tours. Regarding the budget to visit ecotourist sites in Hong Ha, all mentioned that they spent less than USD 50, except for one who spent up to USD 100.

5 Achievements and Challenges of the Ecotourism Programme in Hong Ha

With these initiatives, Hong Ha's ecotourism programme has achieved overwhelming results. Between March 2016 and August 2016, approximately 19,000 tourists visited ecotourism sites in Hong Ha. Meanwhile, approximately 12,000 tourists visited Hong Ha from April to July 2017 (Fig. 23). At the most, there were 562 daily visitors on 30 April and 1 May. From March 2016 to August 2016, the income generated from tourists totalled VND 228,898,000 (approximately USD 10,095). However, the total income within the four months of 2017 (from April to July) was VND 215,416,000 (approximately USD 9,500) (Fig. 24). The average monthly income from the ecotourism project was VND 2,900,000 (approximately USD 128). The cost of purchasing materials was VND 12,300,000 (about USD 542), payments to members working in the ecotourism project amounted to VND 110,756,000 (about USD 4,885) and the management fee to the Hong Ha Commune was 10% of the income, at VND 22,889,800 (about USD 1,010). These data prove the increase in the number of tourists to Hong Ha and the need to develop ecotourism programmes.

The ecotourism programme is useful to the Hong Ha commune in various ways. It generates additional income for local people participating in the programme, while traditional values can be improved and expanded through dance performances and weaving in community houses. Teaching activities in primary schools can effectively conserve traditional values across generations. Hong Ha's natural and human resources can be utilised and popularised as attractive ecotourism destinations.

However, ecotourism in Hong Ha commune has certain challenges. For sustainable development of ecotourism, there should be a balance between humans, natural resources and income (Pham et al., 2002). The expeditious development of ecotourism in Hong Ha has some negative consequences. The construction of roads and infrastructure near Parle waterfall, have negatively affected natural resources.

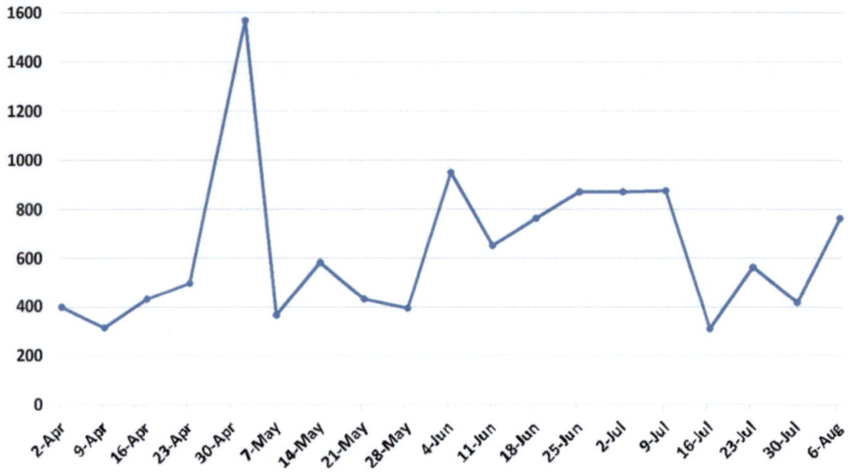

Fig. 23 Number of tourists visiting Hong Ha between April and July 2017

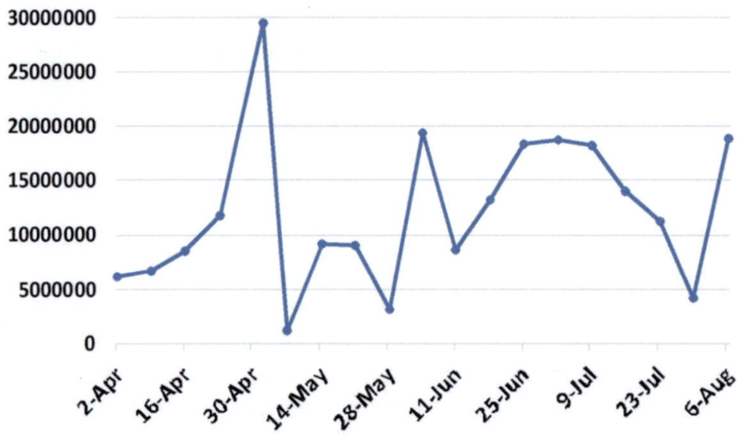

Fig. 24 Total weekly income from ecotourism in Hong Ha from April to July 2017

In addition, a cultural village project supported by the national government was launched in 2017 (Fig. 25). This project includes the construction of a concrete community house and five small community houses representing the five hamlets in Hong Ha. However, this minimises the value of Hong Ha's natural landscape (Nguyen et al., 2019).

In Parle waterfall, one washroom caters to more than 100 tourists. Some tourists dispose of garbage in an irresponsible manner. Currently, ecotourism in Hong Ha is simple and offers limited options for tourists (People's Committee, 2016b). Three sites (Parle waterfall, community house and A Zoi stone) mainly serve as ecotourism sites. In addition, connectivity between tourist sites in Hong Ha and other areas is

Fig. 25 Current status of cultural village projects

poor. Therefore, the lack of adequate washrooms in Parle waterfall, solution to solid waste disposal, disorderly traffic and the lack of connectivity among tourist places in Hong Ha in particular and Thua Thien Hue in general are some of the challenges for ecotourism programmes.

6 Recommendations for Sustainable Initiatives of Hong Ha Ecotourism

The respondents included local people, local government officials and tourists. Based on the survey results, several primary recommendations for ecotourism in Hong Ha have been provided.

- Hong Ha has the potential to develop its ecotourism because the area has various resources, such as Parle waterfall, community house, A Zoi stone, forest, landscape and human resources.
- For future ecotourism projects, Hong Ha should plan to construct more washrooms at Parle waterfall, establish homestays for tourists, construct guesthouses for tourists staying overnight in Hong Ha, and expand more sites such as A Rum stream and forest landscapes.

- Regarding human resources, Hong Ha should establish training courses for programme staff to improve their ecotourism skills and experiences.
- Information about ecotourism in Hong Ha has not yet been popularised. Therefore, Hong Ha should strengthen its advertising and tap resources such as social networks (Facebook and websites), newspapers, television reports and associate with travel agencies.

The following are some solutions for future ecotourism development in Hong Ha.

6.1 Initiative for Various and Attractive Tourist Sites

Hong Ha should be marked as a potential tourist spot (also known as a strategic points). Strategic points are scattered across various places, and by definition, should be redesigned as a system. A public area, such as a traditional community house, should serve as the point of contact to enhance operations in sub-regional areas, instead of being just a self-referential entity. Furthermore, the strategic points that could be linked by road and stream should be identified and enhanced.

Hong Ha can select various potential sites for ecotourism, such as traditional community houses, Parle waterfall, A Zoi stone, landscape along Bo river, forest and natural areas, A Rum stream and Pilo ravine. The commune should create various and vivid tourist programmes in which travellers have many choices when visiting Hong Ha.

6.2 Traditional Community House as a Centre for Ecotourism

Community houses have historical and architectural value. Installing it as a central connection point is crucial. It becomes a visitor centre or a space for interpretation and knowledge. It concerns many objects such as architecture and territory because of its complex history and culture. In a community house, local people and authorities should be trained in ecotourism (in the implementation of guidance services, accommodation, travelling, food, etc.). Additionally, a community house can be used as a place of residence; however, this can also affect visitors during the day. Therefore, it is imperative to build accommodation or use residential houses as homestays.

6.3 Establishment of Tourist Network

These sites comprise a network that contributes towards valorising each other. Tourists find it difficult to approach specific locations. The first point they need to get to is the traditional community house; after obtaining information on the travel programme, they will be able to go to other destinations as directed.

Each strategic site is characterised by different conditions; therefore, it hosts different design themes.

6.4 The Adaptability Architecture in Each Strategy Sites

When an ecotourism system is established, buildings should be set up to support tourists, such as small huts (to rest beside the road) and bungalows (along the stream). It can serve as a node or satellite in a travel network.

The research does not wish for eternal buildings but for structures that follow the seasons and adapt to them. Our installations are modifiable in relation to time and weather conditions, and the passage of months and years will change the appearance and layout of the interventions. The buildings consist of a sequence of local wooden ribs that shape the structure depending on the dimensions and number. It can be considered a small wooden bungalow.

All installations have the same characteristics. Installation is the best method to avoid storms and flooding. The structures will be mainly made of wood and the technology to create them will be taught in the community house to local people and many generations.

7 Conclusion

The case study of Hong Ha commune in Vietnam illustrates a promising model of community-based ecotourism in a rural Southeast Asian context. By mobilising local natural and cultural resources—such as forests, rivers, waterfalls, and traditional community houses—the commune has successfully generated alternative income streams while preserving elements of intangible cultural heritage. A particularly notable feature of this model is its self-managed, locally led system, which reduces dependency on central government support and reinforces community ownership of tourism development.

However, the study also reveals practical challenges that need to be addressed for sustainable ecotourism growth. These include the lack of sufficient washroom facilities at major sites like Parle waterfall, inadequate infrastructure connecting ecotourism locations, and limited visitor management leading to issues such as improper waste disposal. Additionally, some infrastructure development—such as the construction of roads near natural attractions—has unintentionally harmed local ecosystems. These challenges underscore the importance of introducing strategic governance frameworks that can guide ecotourism development while protecting environmental integrity and enhancing the visitor experience.

From a theoretical perspective, this case contributes to the broader literature on community-based tourism and grassroots sustainability. It highlights how small-scale, locally initiated tourism efforts can foster rural revitalisation, environmental stewardship, and participatory governance when grounded in strong local engagement. Such models offer valuable insights for similarly situated communities in Southeast Asia and beyond, especially in contexts where top-down support is limited.

Future research should further explore the long-term impacts of community-led ecotourism initiatives on local economies, natural resources, and cultural practices. Comparative studies across different regional models could provide a deeper understanding of the mechanisms that contribute to resilience and sustainability in rural tourism systems.

Acknowledgements This research was supported by GSGES seeds research funding programme 2016–2017 and CoHHO, Kyoto University, JAPAN. We acknowledge this support.

References

Dang, N. V., Chu., T. S. & Luu, H. (1993). *Ethnic Minorities in Vietnam*. Hanoi, Vietnam: Thế Giới Publishers

Kobayashi, H., Nguyen, X. H., Le, V. A., Ho, T. T., Tran, N. L., Tanaka, U., Mizuno, K., Iizuka, A., Nguyen, T. N. T., & Le, T. H. N. (Eds.). (2008). *Participatory construction of traditional community house in mountainous village of Central Vietnam*. National Political Publisher.

Hue Monuments Conservation Center. (2007). *Report for traveler number and sales from 2010 to 2016*. (In Vietnamese)

Luu, H. (2007). *A Contribution to Katu ethnography*. The Gioi Publishers.

Nguyen, T. H. (2016). Cột đá thiêng Azoi và dấu ấn lịch sử [Azoi holy stone and historic mark]. *Tạp chí Xưa Nay [Journal of Past Present] no., 475*, 55–60. (in Vietnamese).

Nguyen, Q. T., Ubukata, F., & Nguyen, C. D. (2019). Development process of the community-based tourism: From two case studies in central Vietnam. In: *Proceedings of the Vietnam - Japan science and technology symposium (VJST)* (pp. 124–128). Available from: https://csdlkhoahoc.hueuni.edu.vn/data/2021/6/VJST_2019.pdf

People's Committee of Hong Ha commune. (2016a). *Plan for developing culture tourism in Hong Ha commune in the period 2015–2020*. (in Vietnamese).

People's Committee of Hong Ha commune. (2016b). *Tourism report in the first six months of the year in Hong Ha*. (in Vietnamese).

People's Committee of Hong Ha commune. (2017). *Report for implementing tourist development in the first six months of the year 2017 in Hong Ha*. (in Vietnamese).

Pham, T. L. (Ed.), Hoang, H. Q., Nguyen, N. K., Nguyen, V. L., & Do, Q. T. (2002). *Du lịch sinh thái: những vấn đề về lý luận và phát triển thực tiễn ở Việt Nam [Ecotourism in Vietnam: Theory and practice in Vietnam]*. Education Publishing House. (in Vietnam)

Ton, T. H. D. (2014). Integrated potential assessment of community-based ecotourism in district A luoi, Thua Thien Hue Province. *Vietnam Journal of Earth Sciences, 36*(3), 271–280. https://doi.org/10.15625/0866-7187/36/3/5910

Tran, V. L. (2012). *Report on ecotourism in Thua Thien Hue Province, potentials, status and development orientations*. Department of Tourism, Thua Thien Hue. (in Vietnamese)

Tran, V. T. (1984). Dân tộc Cơ Tu (Co tu ethnic group). In Nguyen, Q. L., et al., *Các dân tộc ít người ở Bình Trị Thiên (Ethnic minorities in Binh Tri Thien)* (pp. 183–218). Hue, Vietnam: Thuận Hóa Publishers. (in Vietnamese).

WWF. (2002). *Assessment of the special use forest system and its management in mountainous areas of Thua Thien Hue, Central Vietnam.* Technical Report No. 4. WWF. Available from https://wwfeu.awsassets.panda.org/downloads/report_2.pdf

Community Initiatives Drive Sustainable Resource Utilization Across Germany: Eco-tourism in Bad Hindelang, Environmental Education in Templin, and Power Generation in Schönau im Schwarzwald

Holger Schaefer

Abstract This chapter describes three community initiatives in Germany that utilize local natural resources in markedly different ways. These initiatives involve the maintenance of alpine pastures for eco-tourism in Bad Hindelang, the management of a school forest for environmental education in Templin, and community-based power generation using renewable energies in Schönau im Schwarzwald. The origins, activities, direct and indirect outputs as well as the challenges of each initiative were depicted. Common success factors of these initiatives, such as the choice of a location-based approach or the use of extensive and integrative information campaigns, were identified in the concluding section.

Keywords Community-based resource management · Sustainable tourism · Environmental education · Renewable energy · Local initiatives · Germany

1 Introduction

In Germany, rural communities comprise 60% of the land area and, thereby, administrate a major proportion of the country's natural resources (Federal Bureau of Statistics, 2016). Mining, agriculture, and forestry, however, have lost their economic significance in rural areas. Today, only 1.8% of jobs are provided in the primary economic sector (Federal Institute for Research on Building, Urban Affairs & Spatial Development, 2017). In order to sustain the local livelihood and conquer economic decline, rural communities are therefore challenged to create new modes of resource

H. Schaefer (✉)
Graduate School of Global Environmental Studies, Kyoto University, Kyoto, Japan
e-mail: schaefer.holger.24x@kyoto-u.jp

Department of Forest Soils, Forestry and Forest Products Research Institute, Tsukuba, Japan

M. Yoshizumi et al. (eds.), *Navigating Local Sustainability in Food, Community, and Innovation*, https://doi.org/10.1007/978-981-96-8963-7_9

utilization. This chapter introduces three German communities, where local initiatives have successfully reconsidered resource utilization, in order to cope with major economic changes (Fig. 1). In Bad Hindelang, a farmers' initiative has preserved traditional alpine farming and the associated cultural landscape, withstanding general intensification and mechanization of agriculture. In Templin, the school forest initiative has strengthened the community's role as an education hub, which had been threatened by economic reorganization accompanying the German reunification. In Schönau im Schwarzwald (hereafter referred to as Schönau), a citizen initiative has promoted community-based power generation based on renewable energies to oppose the stagnant centralized energy industry. The three community initiatives were chosen for the following features. Each of the three community initiatives are citizen-based and have largely sustained themselves without financial support from outside of the community. They have lasted for 10 to more than 30 years, so their long-term development is observed. Each community initiative has produced measureable positive output within the community and made a visible impact beyond the community.

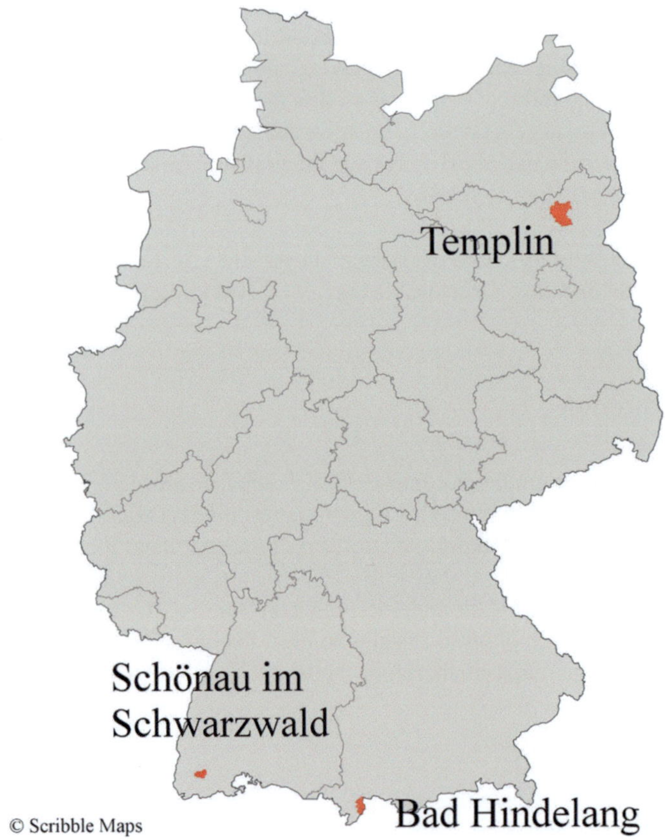

Fig. 1 Map of Germany with the three communities presented in this chapter

The purpose of this chapter is to depict the origins, activities, outcomes, and challenges of each community initiative, and to extract common factors for success and common challenges faced. Information on each of the three community initiatives was collected through a review of scientific articles, books, and informative materials as well as through interviews conducted with respective key experts in December 2017. For the initiative in Bad Hindelang, the community's tourism director Maximilian Hillmeier was interviewed. For the initiative in Templin, school forester Joachim Lange was interviewed. For the initiative in Schönau, CEO of the local power supplier and initiative member Sebastian Sladek was interviewed. If not indicated otherwise, information regarding each initiative stems from the conducted interviews with key experts.

2 Case Study 1: Preservation of Traditional Alpine Farming as a Driver of Eco-tourism in Bad Hindelang

2.1 *Background of the Community*

Bad Hindelang is a market town close to the Germany-Austria border in the federal state of Bavaria (Fig. 1). The community has 5,100 inhabitants living in six villages and six hamlets in the valleys of the rivers Ostrach and Wertach and on the interjacent mountain pass (Community of Bad Hindelang, 2018a). The 13,718 ha-wide community area is part of the Northern Limestone Alps. Elevations within community borders range from 800 to over 2,000 m above sea level, creating a geomorphologically diverse landscape (Bavarian State Office for Statistics, 2018; Community of Bad Hindelang, 2006).

The landscape of Bad Hindelang was largely shaped by the local economy. Between the 15th and early nineteenth century, ore mining and salt trading were the basis of the local economy (Community of Bad Hindelang, 2018a; Lindner, 2000). Already centuries ago, ore mining activities and expansion of agricultural land led to clearance of much of the forest in Bad Hindelang (Lindner, 2000). After mining activities had ceased and salt trading routes had changed, dairy farming became the most important economic pillar in Bad Hindelang during the nineteenth century. So-called alpine farming developed a special kind of dairy farming that efficiently utilizes pastureland in the local mountainous landscape. Thereby, cattle are droved from pastures in valley bottoms to the wide pastureland in higher elevations between June and September. Only in this limited period, local climate allows utilization of the wide elevated pastureland. Already in the thirteenth century Bad Hindelang's farmers drove cattle from the valley to alpine pastures in the *Tannheimer Tal*. Today, Bad Hindelang has 45 alpine pastures run by alpine herdsman in higher elevations, with a total area of about 8,000 ha (Fig. 2; Community of Bad Hindelang, unpublished). In the valley bottoms, about 70 farmers maintain 1,500 ha of pastureland (Department for Food, Agriculture & Forestry Bavaria, unpublished).

Fig. 2 A young herdsman overlooks cattle on an alpine pasture in Bad Hindelang. (Author: Wolfgang B. Kleiner/Bad Hindelang Tourismus)

The pastureland with only small forest patches scattered in between, formed a cultural landscape that contributed largely to a growing tourism in the twentieth century. The number of overnight stays per year increased from 3,500 in 1890 to over one million in 1990 (Community of Bad Hindelang, 1997, 2018a; Lindner, 2000). Also ecologically, the cultural landscape of Bad Hindelang is of high importance. It is the largest connected system of alpine pastures in Germany. The humus-rich shallow soils on the slopes are a habitat for 70 plants on the red list, 40 orchid species and other rare plant species (Community of Bad Hindelang, 2018b; Ringler, 2014). Today more than 80% of the area of Bad Hindelang is under nature or landscape protection (Community of Bad Hindelang, 2006).

2.2 *Evolution of the Problems Surrounding the Community*

Until the 1950s, agriculture in Germany was characterized by intensive manual labor and small farm sizes (Federal Ministry of Food & Agriculture, 2016a). With increasing proliferation of farm tractors, mechanical milkers, and other large farming machines, however, the necessary workforce per unit of agricultural land or live stock decreased significantly. At the same time, financial investments into farming equipment increased (Federal Ministry of Food & Agriculture, 2016a). This development benefited growth-oriented agricultural businesses, while many smaller agricultural businesses were not competitive anymore. Between 1949 and 1990, the number of agricultural businesses in Germany decreased from about 1,800,000 to 700,000, while the average area per agricultural business increased from 7.4 to 17.7 ha (Federal Ministry of Food & Agriculture, 2016b). Since alpine farming is commonly done in difficult terrain, with a limited amount of usable land, steep slopes, and shallow soils, the alpine farming businesses in particular became less competitive (Bund Naturschutz in Bayern e.V., 2004). In Bad Hindelang, the total number of farming businesses decreased drastically from 216 to 73 between 1969 and 1988 (Lindner, 2000; Wirthensohn, 1988). Alpine farming in Bad Hindelang sustained a cultural landscape that has become the main resource for the local tourism and an important habitat for diverse plant communities. With a reduction in alpine farming, pastureland of Bad Hindelang was about to become partially overgrown by shrubs. This was a threat to the community's economic situation and ecological significance.

2.3 Purpose of the Initiative

Farmers of Bad Hindelang founded the *Verein Hindelang Natur und Kultur* (association Hindeland nature and culture) in 1992. The purposes of the farmers' initiative were to (1) maintain local alpine farming practices, (2) market agricultural products locally, and, thereby, (3) preserve the local cultural landscape and its economic and ecological significance.

2.4 The Origin, Evolution, and Current Status of the Initiative

In the 1980s, the German national government and the Bavarian federal state government created several funding programs to support alpine farming. Embracing such funding opportunities, the educational advisor Ernst Wirthensohn of the *Bund Naturschutz in Bayern e.V.* (nature conservation association in Bavaria) developed a strategy paper for the conservation of alpine farming in German communities. The paper explained in detail how farmers could attain the newly available funds through implementation of extensive, organic farming (Lindner, 2000). When introduced to the community of Bad Hindelang, Kasper Weber, farmer and chairman of the *Wald und Weidegenossenschaft Hinterstein* (forest and pasture cooperative Hinterstein), then mayor Roman Haug, and then tourism director Walter Besler decided to implement the strategy. They started the so-called *Ökomodell Hindelang* (eco-model Hindelang).

Eight farmers around Kasper Weber in the village of Hinterstein committed themselves to extensive and organic farming right away. At the same time, large effort was put into advising and convincing farmers of Bad Hindelang who were still critical of the idea. For the promotion the *Ökomodell Hindelang*, Kasper Weber received the German Environmental Award in 1990. Eventually, the initiators' work and the attention they had received nationwide made 86 of 87 farmers of the community participate. As a representative body the farmers founded the initiative *Verein Hindelang Natur und Kultur* at the annual meeting of the International Commission for the Protection of the Alps (CIPRA) in Schwangau, Germany in 1992 (Lindner, 2000).

From then on, the members of the initiative acted as conservators of the community's cultural landscape. For their activities, farmers were provided financial support from several sources. Farmers received about DEM 960 (1992, equivalent to EUR 720 in 2017) per ha land annually from national and federal state funding programs. From the town administration of Bad Hindelang, farmers received 150,000 DEM (1992, equivalent to EUR 113,000 in 2017) annually (Lindner, 2000). A third source of income were private sponsors such as the Japanese copy machine producer RISO who paid DEM 100,000 (1994, equivalent to EUR 70,000 in 2017) for several years, or the Aga Khan Foundation.

The initiative has been successful until today and is acknowledged as the basis for the current tourism strategy of Bad Hindelang (Community of Bad Hindelang, 2017). National and state funding as well as the direct financial support by the town administration have been carried on until today. Currently, the farmers receive EUR 60,000 annually from the community household. In 2016, alpine farming in Bad Hindelang was included as one of four Best Safeguarding Practices into the German Nationwide Inventory of Intangible Cultural Heritage. The inventory is part of the implementation of the UNESCO Convention for the Safeguarding of the Intangible Cultural Heritage from 2003 (German Commission for UNESCO, 2017).

2.5 Description of the Initiative's Activities

The initiative's main activity is the conservation of 1,500 ha pastureland around the villages and hamlets of Bad Hindelang (Fig. 3). Farming principles have been the following: (1) to maintain farming activities all year round, (2) to keep at maximum one unit of cattle (500 kg) per ha pasture land, (3) to produce 90% of the used feed within the community area, (4) to refrain from using inorganic nitrogen fertilizers, (5) to conduct only selective weed control, and (6) to refrain from selling milk allotments to outside of the community (Lindner, 2000). From June to September every year, cattle are droved from the settlements in the valley to 8,000 ha of alpine pastureland in higher elevations (Community of Bad Hindelang, unpublished). These summer pastures are managed by alpine herdsman and other local staff. The work on the summer pastures is not part of the initiative's activities but depends on the provision of cattle by farmers in the valleys.

In order to promote farmer-to-consumer direct marketing, 65 farmers founded the company *Hindelanger Bauern-Vermarktungs GmbH & Co. KG* (Hindelang farmers' marketing company). By 1995, two local alpine dairies had opened, processing and selling locally produced milk (Community of Bad Hindelang, 2006). In addition, a

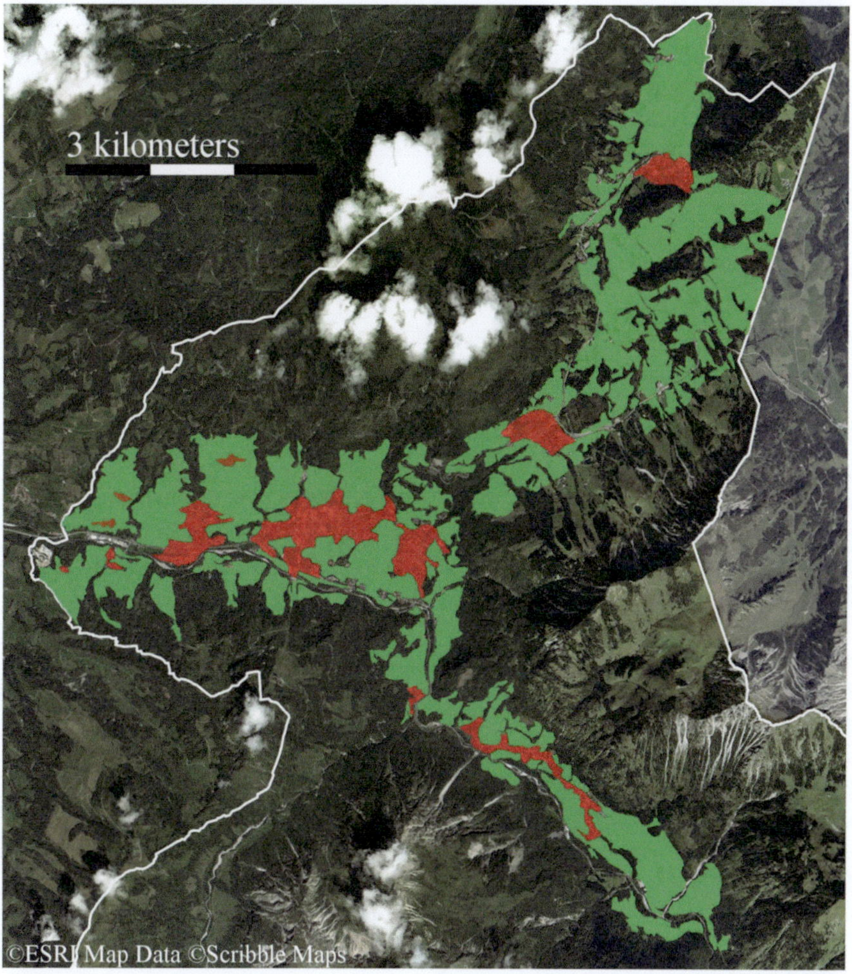

Fig. 3 North-oriented satellite image of Bad Hindelang, showing pastureland (marked green) below an elevation of 1,100 m in the valleys of the rivers Ostrach (southeast to west) and Wertach (center to north east), and below 1,200 m on the mountain pass in between. Villages and hamlets are marked red, community borders white. Marked pastures do not equal the initiative's pastureland, but only indicate their approximate location and area

butcher's shop opened that sold locally produced meat (Federal Agency for Nature Conservation, 2007; Lindner, 2000). Today, agricultural products are, for instance, offered by local caterers, a small farmer's shop, and a regular farmers' and organic market (Schmid, unpublished).

Additionally, the initiative maintains several activities that are directly related to the local eco-tourism. An essential part of traditional alpine farming is the *Viehscheid* in September every year. Thereby, joint herds from Bad Hindelang's farmers are droved back from elevated summer pastures to the valleys. Each head of cattle is returned to its owner. The *Viehscheid* in Bad Hindelang has been a popular tourism event. In 2009, farmers started the farm stay program *Urlaub auf dem Bauernhof* (holiday on the farm) which is gaining popularity. In 2016, there were 21 farmers who hosted rooms or holiday apartments in the community area (Community of Bad Hindelang, 2017).

2.6 Output and Outcome of the Activities

The number of farmers in Bad Hindelang decreased from 87 to 67 between 1992 and 2013 (Schmid, unpublished). The area of pastureland maintained by the initiative, however, has barely changed. There were 1,548 ha of pastureland in 1993 and 1,564 ha in 2017 (Department for Food, Agriculture & Forestry Bavaria, unpublished; Schmid, unpublished). This is explained by the high motivation of the initiative's members. Farmers that retired from their businesses continued managing their pastureland as a hobby. Pastureland that was given up on was taken over by neighboring farmers. The cultural landscape around the settlements of Bad Hindelang has therefore been conserved successfully.

Directly related to the conserved cultural landscape is the tourism in Bad Hindelang. The number of visitors per year has continuously increased from 110,000 in 1993 to 225,000 in 2016 (Community of Bad Hindelang, 2014, 2017). Although the community is within a popular ski-region, the number of visitors between July and September has been as high as during the ski-season from January to March (Community of Bad Hindelang, 2017). Just for the annual *Viehscheid* in September, Bad Hindelang has had about 20,000 visitors (Fig. 4; Horn, 2013). The farm stay program of the initiative's farmers, *Urlaub auf dem Bauernhof* has attracted about 5,000 additional visitors per year (Community of Bad Hindelang, 2017). Bad Hindelang's community administration described the initiative's activities as integral for its ongoing success in the tourism sector and for its future tourism strategy (Community of Bad Hindelang, 2017).

The farmer-to-consumer direct marketing was quickly implemented in the early 1990s and has been popular with tourists and locals. The two local alpine dairies have been processing and marketing approximately 40% of the locally produced milk (Community of Bad Hindelang, 2006). The local butcher's shop has been selling around 90% of locally produced meat (Lindner, 2000). In total, 35 cows, 150 pigs, 20 lambs, 10 deer, 300 kg cheese were marketed in the butcher's shop, the two alpine dairies, and five local food caterers in 2001 (Federal Agency for Nature Conservation, 2007). Through the direct marketing, farmers of the initiative have been able to sell their quality products for increased prices and effectively support their livelihood.

The initiative may also be credited for its educational achievement, as it transferred traditional farming skills into a modern economic context. Agricultural products of the initiative's farmers have satisfied standards of major organic labels in Germany as *Bioland*, *Naturland*, or *Demeter*. Already in 2002, these labels had officially certified the products of 20 farmers (Community of Bad Hindelang, 2006). However, not only

Fig. 4 The traditional *Viehscheid* in Bad Hindelang is a popular tourist attraction. (Author: Wolfgang B. Kleiner/Bad Hindelang Tourismus)

the traditional skills and knowledge of the initiative's farmers have been preserved this way. Droving more than 200 dairy cows and 2,500 young cattle to elevated summer pastures every year helps to sustain activities of the local alpine herdsman (Community of Bad Hindelang, unpublished).

Beyond the community-scale, the initiative of Bad Hindelang has become a model for successful conservation and utilization of a cultural landscape. Since its start, the initiative has received many prizes, was the focus of scientific publications, and recently appeared in a geography schoolbook for secondary school students. The initiative has therefore largely contributed to raising awareness regarding the ecological and economical importance of alpine farming specifically and cultural landscapes generally.

2.7 Lessons Learned and Challenges Faced by the Initiative

The initiative in Bad Hindelang was challenged when implementing several direct-marketing concepts. From early on, farmers of the initiative attempted to cooperate with local caterers to market their products. The idea was that restaurants and hotels offer dishes made from the initiative's products, specifically addressing customers interested in regional products. These dishes were highlighted on menus by a special slogan and had a circa 20% higher price (Lindner, 2000). However, many local caterers were not willing to pay an increased price to attain the initiative's products, as farmers were receiving large financial support from several sources. By 2001, only five local caterers had cooperated with the initiative (Federal Agency for Nature Conservation, 2007).

In 2001, the initiative opened a central farmers' market along a major road within the community area. The farmers' market was a large investment for the initiative's farmers who beard much of the costs of EUR 4.5 million themselves (Federal Agency for Nature Conservation, 2007). Within only eight months, however, the farmers' market went bankrupt, as customer numbers remained low. The initiative's farmers lost a major part of the money they had invested, which was a setback that also slowed down the initiative's activities for several years. One reason for the failure of the project was that local citizens were apparently not willing to pay increased prices for the products sold at the farmers' market (Schmid, unpublished). In addition, marketing products centrally may not have been a promising concept for the initiative in the first place. In contrast to the central farmers' market, decentralized units as the two diaries have been working well.

2.8 Recommendations to Further Improve the Initiative

Currently, the initiative and the community administration are working on new concepts to market the initiative's products in cooperation with other local businesses. For instance, they are planning a breakfast basket for visitors which contains a range of local products, including cheese from the initiative's farmers but also bread from local bakeries etc.

Since the start of the initiative, the number of participating farmers decreased from 86 to around 70. Although the initiative has largely improved the financial situation and esteem of local farmers, traditional farming is hard work and may lack competitiveness against other types of work. In addition, local farms are mostly family businesses and private property, which makes integration of non-local farmers difficult. Tourism director Hillmeier reports little problems with finding successors for the farming businesses in general. Just recently, a young farmer had opened a new farm. Nevertheless, the initiative may need to tackle the issue of how to continue the farmers' landscape conservation with future generations.

3 Case Study 2: Community Forest Becomes a Hub for Practical Environmental Education in Templin

3.1 Background of the Community

Templin is located approximately 80 km from the German capital of Berlin (Fig. 1). Its borders lie within the rural district Uckermark in the federal State of Brandenburg. Templin has an area of 379.5 km^2, with approximately 16,000 inhabitants. This makes Templin one of the least-populated regions in Germany (City of Templin, 2017a). Nevertheless, the city is both a highly significant tourist destination and a regional educational hub.

The tourism sustained by Templin's townscape and surrounding landscape has served as one of the area's important economic pillars. The city hosted 345,602 overnight stays in 2015, which is 45% of all overnight stays in the Uckermark district (City of Templin, 2017a). The structure of Templin's city centre reflects its medieval history. It contains numerous traditional half-timber houses within a densely developed perimeter, surrounded by a 1.7 km-long town wall that was constructed in the fourteenth century (City of Templin, 2017a). Around the city centre, green spaces, forests, and numerous lakes facilitate a thriving eco-tourism. The surrounding landscape was largely shaped by glacial advances and periglacial geomorphologic processes that occurred during the Pleistocene era (Hofmann & Pommer, 2013). With over 33,000 km of rivers and streams, and nearly 10,000 lakes, Templin's aquatic composition makes Brandenburg the most water-rich federal state in Germany (Fig. 5; Hupfer & Nixdorf, 2011). Along with Templin's 3,456 ha community forest, there are several easily accessible natural preservation areas, the Naturpark Uckermärkische Seen, and the Schorfheide-Chorin Biosphere Reserve (City of Templin, 2017a). With such a rich landscape surrounding its city centre, Templin was officially appointed a 'climatic health resort' in 1888, a 'state-approved climatic health resort' in 1971, and a 'state-approved resort town' in 1985 and 1994 (City of Templin, 2017b).

Templin also functions as a regional hub of education. From 1817 to 1993, Templin served as a district city, and housed numerous administrative institutions. During that time, numerous schools were opened to meet educational needs (City of Templin, 2017b). Arguably, the most prestigious of these schools was the now-closed Joachimsthalschen Gymnasium, the history of which began in 1607 (City of Templin, 2017a). Today, Templin contains 12 nursery schools, four primary schools, two secondary schools, a special-needs school, a comprehensive school, and a secondary education centre. Many of these schools are independently operated or utilize a special integrative profile. For example, the Waldhof School (*Waldhofschule Templin*) focuses on special-needs education, but enrolls 260 students both with and without disabilities (Waldhofschule Templin, 2018).

Fig. 5 View of Templin's small port at the *Stadtsee* (Author: Reinhard Kraasch, CC BY-SA 4.0 [https://creativecommons.org/licenses/by-sa/4.0], via Wikimedia Commons)

3.2 Evolution of the Problems Surrounding the Community

Economic reorganizations accompanying the reunification of the German Democratic Republic (East Germany) with the Federal Republic of Germany, represented a large challenge to the community of Templin. The lack of larger industrial settlements made it difficult for a strong private corporate sector to develop (City of Templin, 2017a). Templin also lost its district city status in 1993, and thereby several administrative institutions. In the early 1990s, the unemployment rate was around 12% (Makowitz et al., 2014). With the younger population leaving the city for jobs or training opportunities, out-migration continuously exceeded in-migration for the years following (City of Templin, 2017a). In total, the population of Templin decreased by 10%, going from 19,100 inhabitants in December 1990 to 17,100 inhabitants in December 2006 (City of Templin, 2017a). To adjust to the out-migration of younger generations, the community administration had to close several primary schools and one secondary school (City of Templin, 2017a).

In such a situation, Templin regained the right to manage its 3,456 ha-wide community forest autonomously. For previous decades, East German state institutions had managed the forest (Manig & Verch, 2004). Wood prices in the early 1990s were, however, comparatively low, and the personnel cost for three foresters and several forest workers was high. Such community forest management led to deficits for the city for several years. When the city and forest administrations discussed the necessary measures to make forest management more cost-effective in the early

2000s, drastic solutions such as selling the community forest were even seen as options. The community forest of Templin, however, did not solely provide wood for sale; there were numerous expectations for forest management to maintain the forest as a resource for local eco-tourism.

3.3 Purpose of the School Forest Initiative

In 2007, the communal forest administration, the community administration, and the Waldhof School (*Waldhofschule Templin*) with its operator, the Stephanus Foundation (*Stephanus Stiftung*), implemented a unique concept that combined the sustainable management of Templin's community forest with the practical environmental education of youths and children. Thus, around 700 of the community forest's 3,456 ha were leased to the Waldhof School, which subsequently managed the leased portion of the forest with its students (Fig. 6). The main purposes of the school forest initiative were as follows: (1) to maintain and preserve the community forest through cost-effective forest management, (2) to facilitate the practical study of sustainable natural resource utilization for children, and (3) to involve citizens in the community's forest resource management.

3.4 Origin, Evolution, and Current Status of the School Forest Initiative

The school forest initiative developed into its current shape during negotiations between the forest administration, the community administration, and the Waldhof School, along with its operator, the Stephanus Foundation. Joachim Lange, one of the community's three foresters, devised the basic concept of involving school classes into the management of the community forest. The forest administration was concerned about the lack of citizens' involvement in the management of the community forest, despite the important roles it played in Templin's financial issues, tourism industry, and environmental conservation efforts. The forest administration suggested the concept to the community administration, which welcomed the idea under the condition that it did not induce any costs or lead to a reduction in profits from the community forest. The Waldhof School and its operator were approached later on. Fortunately, the school had already been searching for ways to integrate a business into their school curriculum. It wanted to give students the opportunity to become involved in the daily operations of a company. Eventually, all three parties agreed that the community administration would lease 700 ha of the community forest to the Waldhof School and its operator based on a leasing contract, the details of which were elaborated by the forest administration. Through the leasing contract, the Waldhof School was enabled to independently manage the 700 ha forest area until

Fig. 6 North-oriented satellite image of Templin, showing the location of the school forest (marked green) and the Waldhof School (framed red dot). Urbanized areas are marked red, community borders white

2031, but was required to pay an annual leasing fee. This fee was at least commensurate with the potential profits as speculated by the previous management procedures of the same area. That is, it was required to match the profits received by 4–7 m^3 of harvested wood per ha per year. A key feature of the agreement was that Joachim Lange was required to leave his position as a forester to become a full-time teacher at the Waldhof School following a teacher training program. Thereby, the community administration was able save personnel costs for one forester, while profits from the community forest remained unchanged. School forester Joachim Lange emphasized the trustful relationship that enabled the agreement.

The initiative was successful, and celebrated its 10th anniversary in 2017. In 2016, Joachim Lange received the teacher prize of the State of Brandenburg from the state's governor and minister of education for his engagement in the initiative (Jeutner, 2016). Information about the school forest initiative was also included in a brochure on future-oriented community forest projects by the German Nature and Biodiversity Conservation Union (Naturschutzbund Deutschland e.V., 2013).

3.5 Description of the Activities of the Initiative

The main activity of the school forest initiative has been forest management, conducted by school forester Joachim Lange along with students of the Waldhof School. The initiative has additionally been cooperating with student groups from other local schools. Joachim Lange has a regular weekly teaching schedule (Table 1). The contents of the work in the school forest are determined by currently necessary forest operations as well as organizational aspects. The size of the involved student groups has been varying widely. When large groups of around 20 students required supervision, Joachim Lange was joined by additional teaching staff. The students have been involved in every aspect of forest management, such as tree planting, wood sales, accounting work for expense-budgeting, or the calculation of leasing fees. It has been essential for the initiative that students actively conduct forest management, meaning that any important decision regarding forest management is made and implemented by students. Only for tasks such as tree-felling or transportation, private service providers have been employed. The Waldhof School website provides an example of a lesson at the school forest: *The technology work group removes shrubs from the tractor driveways to allow transportation of wood from the felling location. It was hard work using saws and scissors. Forest service provider Jens Richter once gave on site support and brought his skidder[, a heavy vehicle for transportation of cut logs, HS] to check, if it fits. […] A second work group takes care of the selection and marking of „biotope trees". Thereby they are supported by the bat expert Günther Heise and forest service provider Peter Specht. Much valuable knowledge could thereby be imparted to the students. How does one recognize the dens of the Daubenton's bat, which wood pecker species creates the most dens? Which fungi are rare and which ones decompose dead wood until it becomes soil again? The work group wood harvest selects the trees that are cut this autumn. Prices for straight, thick trees are too bad for selling at the moment.[…] The Waldhof students were therefore greedy in selecting high quality wood. Instead, low quality wood as crooked trees or trees with many branches fetch prices that justify felling* (Waldhofschule Templin, 2018).

3.6 Output and Outcome of the Initiative's Activities

Quantitative outputs that were directly produced by the school forest initiative's activities can be broadly divided into educational, economic, and ecological categories.

From an educational perspective, the school forest functions as an effective classroom for environmental education. About 250 students have been working at the school forest over the course of a given year. Since the start of the initiative in 2007, more than 1,000 total students have been trained in the school forest. Regardless of age, Joachim Lange has received positive feedback from students. These have

Table 1 Weekly schedule for student groups in the school forest

Time	Monday	Tuesday	Wednesday	Thursday	Friday
08.00–08.30			Waldhof school students with special educational needs *(up to 7 students)*	Waldhof school students with special educational needs *(up to 7 students)*	
08.30–09.30	Kinder-garden				Secondary school 10th grade
09.30–11.30		Secondary school 10th grade			
11.30–13.00		Waldhof school 4th grade			Waldhof school 5th/6th grade
13.00–15.00					

participated eagerly in the forest management, and many have been fascinated by the related work and machinery (Fig. 7).

Economically, the initiative has been beneficial for both the community administration and the Waldhof School. To date, the school has paid annual leasing fees of EUR 88,000 on average to the community administration. Furthermore, the community administration has saved annual personnel costs of about EUR 50,000 by removing one of the forester positions. The remaining profits from forest management have been used at the school, for transportation to the school forest, and to buy

Fig. 7 School forest students inspect a forwarder, a vehicle for felled logs (Author: Joachim Lange/ Waldhofschule)

work clothing and tools for students. Additionally, participating school classes have received, in total, EUR 2,000 to 5,000 per school term into their class funds, which are commonly used for school trips and other activities.

Lastly, the initiative has promoted the sustainable development of the community's forest resources. In about 20% of the school forest area, the saplings of broad-leaved tree species have been planted to transform current pine monocultures into mixed forests, which is critical from an ecological and sustainable management perspective. For planting operations and the building of forest roads, EUR 330,000 have been drawn from a common fund operated by the federal State of Brandenburg, the national government of Germany, and the European Union that is available to any forest owner. Numerous biotope trees have been marked and protected from felling to preserve habitats for rare animal species, such as the white-tailed sea eagle. The management of the school forest has also received Forest Stewardship Council (FSC) certification.

In addition, the initiative's outputs involve indirect influences at the community level and beyond. First, the school forest has been functioning as a social hub that promotes interaction among students of different ages (e.g., from kindergarten to secondary school), forest service providers, children of the local children's homes, and members of the local sheltered workshop. The parents of participating students were also informed about the management of the school forest through the stories of their children, and discussed the information with teachers at the Waldhof School. Second, the school forest initiative has represented a flagship project for Templin. The forest school initiative was a pioneering educational project that has triggered a significant amount of interest in and beyond the community. It has thus promoted the image of Templin as a 'school city' that houses innovative educational institutions, and also as a 'resort town' that sustainably manages its forests with positive effects on lakes, ground water, soils, and climate. Thereby the initiative has contributed to multiple goals of the city's Integrated City Development Concept 2030 (City of Templin, 2017a).

3.7 Lessons Learned and Challenges Faced by the Initiative

One of the main challenges during the beginning stages of the initiative was the implementation of the (thus far) only theoretical educational concept. This is because information on projects that integrated a school forest into their school curriculum was then unavailable. For instance, it was unclear how to address students with high special educational needs who may also lack the support of a family. It was also unclear which forest management-related topics were suitable for students of certain ages. Eventually, many of the students with high special educational needs have become the most engaged and perseverant students within the initiative. This was apparently due to a provision in the framework that assured them of the need for their assistance in the initiative and noted the importance of their actions. Regarding the students' ages, the original plan was to cover forest management-related topics with students older than 12 years of age, while younger students would be taught about animal and plant species in the forest. Joachim Lange noticed, however, that younger students of the 3rd and 4th grade had already developed great interest for forest management-related topics, when they were properly prepared. Today, all topics are covered regardless of student age. For instance, forestry machines are shown to kindergarten children, and wood is sold with the help of primary school students.

3.8 Recommendations to Further Improve the Initiative

School forester Joachim Lange would like to expand the school forest initiative to provide long-term work perspectives and job opportunities in the forest to participating students, especially those with special educational needs. Many of the students show great interest in forest management careers. To address this, the Waldhof School wishes to implement further activities, such as firewood sales to private residents in Berlin, game meat sales, the planting and management of fruit tree meadows, and the production of apple juice.

4 Case Study 3: Community-based Power Generation in Schönau im Schwarzwald

4.1 Background of the Community

Schönau im Schwarzwald (hereafter referred to as Schönau) is a small town with 2,450 inhabitants in the federal state of Baden-Württemberg (Fig. 1; State Office for Statistics Baden-Württemberg, 2018). It lies within the Southern Black Forest, the highest section of the 160 km-long mountain range Black Forest in the South German Scarplands (Naturpark Südschwarzwald e. V., 2018). 75% of the 1,472 ha-wide community is forested area on elevations up to 1,300 m above sea level (CAU Schönau, 2017; State Office for Statistics Baden-Württemberg, 2018). The community's residential and commercial areas are situated along the river Wiese on elevations from 540 to 600 m above sea level (Fig. 8; CAU Schönau, 2018a).

Schönau was first mentioned in a document in 1113 as a settlement of the Saint Blaise Abbey (CAU Schönau, 2018b). Between the 13th and 15th century, silver mining led to temporary economic prosperity in the community. From the 18th century onwards, the production of textiles and, later, brushes have largely dominated Schönau's economy (CAU Schönau, 2018b). Today, small and medium-sized manufacturing businesses characterize the local economy (Graichen, 2003). With about 60,000 annual visitors, the 'state-approved climatic health resort' Schönau

Fig. 8 View of the city center of Schönau im Schwarzwald (Author: PantaRhei, CC BY-SA 4.0 [https://creativecommons.org/licenses/by-sa/4.0], via Wikimedia Commons, modified)

also sustains a considerable amount of health, eco-, and winter tourism (Graichen, 2003).

Since 1971, Schönau has formed an administrative union with eight surrounding communities. Schönau hosts numerous public facilities such as a primary and secondary school, a district court, a hospital, and a swimming pool. As such, it provides important infrastructure to the surrounding, more rural communities (CAU Schönau, 2018b; Graichen, 2003). Schönau also used to operate its own power utility that supplied the town with electricity from two hydroelectric power plants. One power plant went into operation as soon as 1905, the other one in 1930 (Janzing et al., 2008). Due to increasing power consumption and demand for enhanced stability in the power supply, however, the town administration decided to sell the local power grid to the regional power supplier *Kraftübertragungswerke Rheinfelden* (KWR) in 1974 and grant the KWR the concession for the community's power supply from then on (Janzing et al., 2008).

4.2 Evolution of the Problems Surrounding the Community

As in Schönau, many community-run power utilities went out of operation during the twentieth century. From the 1920s to the 1990s, the number of power utilities in Germany decreased from about 16,000 to 900. Eventually, only four large corporations generated and distributed 80% of the power nationwide (Graichen, 2003; Schiffer, 1999). This increasingly monopolized energy industry, beyond the sphere of influence by citizens, turned out to be problematic, when it advanced the development of nuclear power plants in the 1960s. Soon after the first nuclear reactors had been built, citizen initiatives across Germany protested against the generation of nuclear power due to safety and health issues. Protests had their first large success in Wyhl, where the construction of a nuclear power plant was stopped in 1977 (Radkau, 2011). The nationwide advancement of nuclear power generation and utilization, however, could not be prevented. Also the KWR, the power supplier of Schönau, which had previously relied solely on hydropower generation, included a nuclear power share of 35% in its power mix by the mid-90 s (Graichen, 2003). The catastrophic nuclear accident in Chernobyl in April 1986 first raised concerns about the risks of nuclear power among many citizens in Schönau. Despite the large distance to the nuclear accident, the Federal Environmental Agency measured a doubling of environmental radiation just 15 km north of Schönau within a week after the nuclear accident (Janzing et al., 2008).

4.3 Purpose of the Initiative

In 1986, citizens of Schönau founded the initiative *Eltern für eine atomfreie Zukunft* (parents for a nuclear-free future) that soon developed to a politically active environmental organization, and, in 1997, to a community-based power supplier. The purposes of the initiative were (1) to urge the national government to create a legal framework that promotes energy-saving and nuclear-free, decentralized power supply, and (2) to withdraw from nuclear power utilization on a community-scale.

4.4 Origin, Evolution and Current Status of the Initiative

After the nuclear accident of Chernobyl in 1986, a group of concerned parents in Schönau started to meet regularly and exchange information regarding the availability of uncontaminated food, etc. (Janzing et al., 2008). Soon the group founded the initiative *Eltern für eine atomfreie Zukunft* and became politically active. The initiative began to organize local events to promote energy-saving and environmentally friendly power generation. Key events as community-wide energy-saving contests appeared in nationwide print media, which also led to attention beyond the regional scale (Graichen, 2003). Most prominent members of the initiative were the trained teacher and homemaker Ursula Sladek and her husband, a physician Michael Sladek, who have remained active promoters of the initiative until today (Janzing et al., 2008).

To withdraw from the use of nuclear power in Schönau, the initiative first attempted to cooperate with the local power supplier KWR. When the KWR refused any cooperation, the initiative saw its only chance to reach their goal in purchasing the local power grid and becoming a power supplier itself (Graichen, 2003). The contract that granted the KWR the concession for the local power supply was about to end in 1994. A majority of the community council, however, was critical of the idea of granting the concession to the initiative instead of the KWR. For a foreseeable increase in the annual concession levy, the community council even decided in 1990 to prepone the renewal of the existing contract with the KWR. In 1991, the initiative successfully refuted this decision by organizing and winning a community-wide referendum (Graichen, 2003). In order to be able to compete with the KWR for a follow-up contract, the initiative founded the company *Elektrizitätswerke Schönau GmbH* (power utility Schönau, limited liability company) in January 1994. Its only shareholder was the *Netzkauf Schönau GbR* (grid purchase Schönau, company constituted under civil law) that was shared by more than 650 local citizens (EWS, 2018). In the meanwhile, elections had changed the majority situation of the community council in favor of the *Elektrizitätswerke Schönau GmbH* (EWS). In 1995, the concession for the local power grid was granted to the EWS. A second community-wide referendum organized by opponents of the EWS in 1996 to refute this decision failed. In addition to gaining the concession, the initiative had to purchase the local power grid from the KWR. The KWR first quoted a price of DEM 8.5 million (1996, equivalent to EUR 5.8 million in 2017) but eventually reduced it to 5.7 million (1997, equivalent to EUR 3.9 million in 2017), in fear of a lawsuit regarding an overstated purchase price. The initiative paid the quoted price with financial assets of participating citizens, an energy fund by the GLS Bank, and money fund-raised in the largely successful nationwide campaign *Ich bin ein Störfall* (I am a disruptive incident). In 1997, the EWS became the power supplier of Schönau. In 1999, the German energy market was liberalized, which abolished the commitment of end-consumers to their regional power supplier. In the following, the EWS began to supply customers with power nationwide.

4.5 Description of the Initiative's Activities

In its early years, the initiative conducted mainly educational activities. For instance, a community-wide energy-saving competition was held several times from 1988. Thereby, local households and institutions competed for the highest energy reduction in a given year. Furthermore, the initiative organized six conferences on renewable energy and energy economic law from 1987 to 1994 (Graichen, 2003). From 1992 to the present day, the initiative has been holding annual multi-day workshops called *Schönauer Stromseminare* (Schönau power workshops) that address many aspects of renewable energy utilization.

As the local power supplier EWS, the initiative continued to promote renewable energy and energy-saving. In order to encourage customers to generate power through renewable energy, the EWS offered free consulting and planning for solar power systems as well as cogeneration systems that couple power generation with heat generation. It launched programs to subsidize the installation of such systems and made it free of charge to feed self-generated power into the local power grid. In order to promote energy-saving, the EWS changed to a linear power tariff system in 1999 (Graichen, 2003). The new system had low base rates but relatively high tariffs for actually consumed power, to reward low-energy consumption.

Within the last 10 years, the EWS has started to generate power by itself. The EWS has built solar power systems in Schönau and other areas. The location of Schönau's residential and commercial areas in a valley bottom have, however, limited the generation of solar power. Therefore, the EWS has more recently invested in the generation of wind power as well as heat generation through biomass combustion. For the latter, wood residues are sourced in cooperation with local communities and private forest owners. Heat from biomass combustion feeds several neighborhood-scale heat grids.

4.6 Output and Outcome of the Activities

The initiative's educational activities mobilized a larger part of the local population from early on. The energy-saving competitions, for instance, animated more than 140 private household and several community-run facilities as local schools to participate. On average, participants reduced their power consumption by 20% (Graichen, 2003). Also the promotion of power generation among Schönau's citizens has been largely successful (Fig. 9). In 1998, the ratio of generated solar power to the total amount of consumed power was 20 times as high as in other German communities. For several years, Schönau had the highest amount of solar power generated per capita in Germany. The solar power system on the local evangelic church, thereby, has been of especially high symbolic importance (Fig. 10). In addition, the EWS itself has generated power. Only within a radius of 10 km around Schönau, the EWS is operating five solar power plants, seven cogeneration plants, five wind turbines as well as two biomass combustion plants that feed a heat grid (Table 2).

In addition, the EWS has become an important employer and tax payer in Schönau. When the EWS took over the local power grid, it had one paid employee, while the other staff worked voluntarily. Today, the EWS has about 130 paid employees, of which more than 90% are from the local area. With a sales volume of more than EUR 3 million in 2016 (EWS, 2017), the EWS has become the largest payer of commercial tax in Schönau. As the community administration of Schönau has been rather limited in its capacity to invest locally (Graichen, 2003), tax payments have helped to improve local infrastructure such as the recently renovated public swimming pool. The EWS has also financially supported local clubs and societies, as the carnival association. Annual events such as the carnival festival and the Christmas market have been provided with power free of charge.

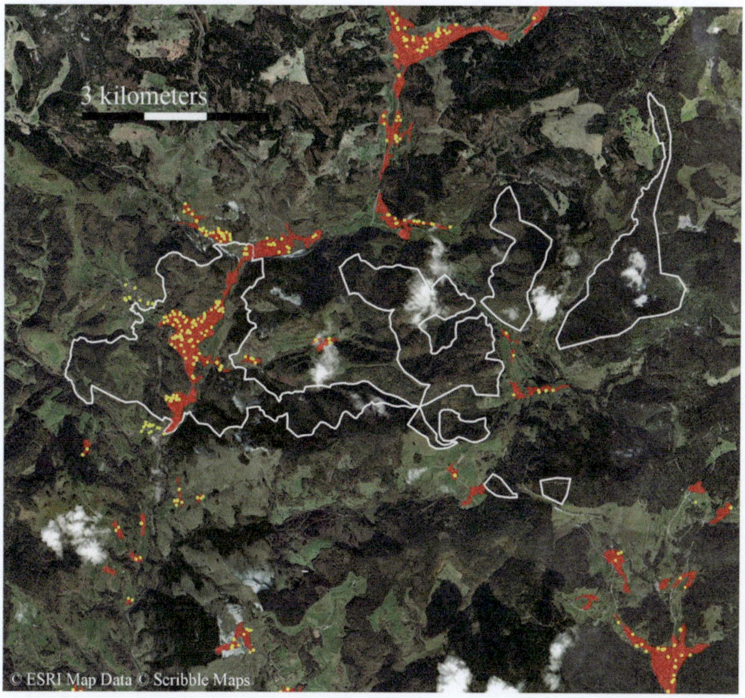

Fig. 9 North-oriented satellite image of Schönau, showing the location of solar power systems clearly visible on satellite images (marked yellow). Urbanized areas are marked red, community borders white

The initiative has largely contributed to popularizing the utilization of renewable energies as well as community-based power generation. Within the community, many identify with the successful transformation of the local power supply. Beyond the community, the initiative's fight for the right to operate the local power grid has attracted much attention in the media. The nationwide fund-raising campaign *Ich bin ein Störfall* in 1996 gained the support of thousands of citizens and all major environmental associations in Germany. Numerous politicians and renowned scientists have visited the community (Janzing et al., 2008). In 2002, EWS customers nationwide provided about 20% of the total installed performance of solar power systems in Germany. Today, there are 2,400 citizen-owned solar power systems subsidized by the EWS (EWS, 2018). The attention the initiative attracted has resulted in a type of energy tourism. The yearly workshops *Schönauer Stromseminare* are attended by visitors from all over the world. Schönau even created the energy experience park *Energieerlebnispark Buchenbrändle* that teaches about different types of power generation and their relation to climate change (CAU Schönau, 2018b).

Fig. 10 Solar power system on the *Bergkirche*, a local protestant church (Author: Wladyslaw Sojka, www.sojka.photo, via Wikimedia Commons)

Table 2 Power generation sites owned by the EWS within a radius of 10 km around Schönau

Type	Number of sites	Location	Total electric capacity (kW)	Total thermal capacity (kW)
Solar	5	Schönau	97*	–
Cogeneration	7	Schönau, Zell, Wembach	112	233
Biomass	2	Zell	-	1,700
Wind	5	Gersbach	15.000	–

Information based on https://www.ews-schoenau.de/ews/energieerzeugung/energieerzeugungsanlagen/.
*For solar power system, the electric capacity equals the electric capacity peak (kWp).

4.7 Lessons Learned and Challenges Faced by the Initiatives

The largest challenge for the initiative was the local political conflict surrounding the local power supply. In the beginning, neither the community administration nor the local power supplier KWR wanted to cooperate with the initiative. A large part of the initiative's work in the early 1990s, therefore, consisted of locally campaigning for their idea of community-based power generation. Information politics and integrative events as the energy-saving contests were essential to mobilize a large part of the local population and win two community-wide referenda. Also today, the EWS is eager to involve citizens into their activities as soon as possible. The wind park *Rohrenkopf*, for instance, had 1.5 years of forerun and was discussed multiple times in local community councils.

A second challenge to the initiative has been to balance economic decisions of a power supplier with its philosophical ideas. As not to concentrate all power related tasks (grid operation, generation, supply) on itself and become a type of energy monopolist, the EWS avoided generating power by itself for many years. For further promotion of renewable energy utilization, however, it became inevitable to build larger power generation sites such as wind parks.

4.8 Recommendations to Further Improve the Initiative

While the EWS is advancing its construction of power generation sites locally, the initiative in Schönau plans to become more politically active from 2018. Energy politics in Germany have become increasingly restrictive. Many similar initiatives in other communities have failed to become community-based power suppliers, as they are largely judged by their economic competitiveness. Even if they gain sufficient financial funds and a solid basis of supporters, initiatives have barely had a chance against large-scale power suppliers. Therefore, the initiative in Schönau has been increasingly acting as a supporter of potential community-based power suppliers in Germany. Due to similarly difficult circumstances for potential community-based power suppliers in other countries, these activities have recently extended beyond national borders. The EWS is, for instance, supporting an initiative on the small French island Île de Sein that wants to switch to renewable energy.

5 Conclusion

In each of the three communities, the initiative is directly linked to a history of resource utilization beyond the production of raw materials in mining, agriculture, and forestry. In Bad Hindelang, this link is most obvious, as the farmers' initiative has maintained the already established landscape-based eco-tourism. In Templin, the creation of a school forest relates to a history of community-based forest management, which has served not only the production of wood but also secondary functions such as the landscape's recreational service. In Schönau, citizens have had utilized renewable energy, in the form of hydropower, for an autonomous power supply as early as 1905. Hence, either of the initiatives is location-based, not just in the sense that it relates to available natural resources but also to the way they have been utilized.

This instance may have contributed to the successful implementation of community initiatives, as promoted ideas were accepted more easily by the local population.

A second key issue impeding the eventual success of initiatives is the initial constellation of stakeholders. In Templin, the city administration, the forest administration, and the Waldhof School maintained a trustful relationship and soon agreed on a concept for the school forest that was profitable for every party. In Bad Hindelang, individual farmers and the community administration were agreeing on the necessity of a farmer's initiative, but many more farmers needed to be convinced for successful landscape conservation. In Schönau, the initiative's intention to become the local power supplier was opposed by the community administration as well as the previous power supplier KWR. This initial constellation of stakeholders, made the initiative in Templin develop rather smoothly, while initiatives in Bad Hindelang and Schönau needed to overcome larger initial hurdles.

The main tool used by initiatives in Bad Hindelang and Schönau to overcome initial hurdles were extensive information campaigns. In Bad Hindelang, this was done by advising so far critical farmers on how they can profit from altering their farming practices. In Schönau, information campaigns on the benefits of energy-saving and the withdrawal from nuclear energy directly addressed the local population. In addition to providing information, the potential of altered resource utilization was visualized. In Bad Hindelang, eight farmers around Kasper Weber implemented advertized farming practices right away. In Schönau, the initiative had used integrative events such as community-wide energy-saving competitions from early on, allowing citizens to grasp their impact on the local power consumption. These campaigns eventually created a common understanding of how local resources need to be managed.

Over the long term, the flexibility of community initiatives has contributed to their eventual success. In each of the communities, activities of the initiative have changed over time. In Bad Hindelang, farmers started, for instance, a farm stay program in 2009, after the success of cooperation with local caterers had been limited and the central farmers' market failing completely. In Templin, lesson contents in the school forest have changed over time. After early attempts to choose age-appropriate lesson topics, all aspects of forest management are now covered regardless of the students' age. In Schönau, the initiative changed from a politically active organization over a local power supplier to a power generator and nationwide power supplier. All initiatives adjusted themselves to changing circumstances, slowly developing towards their current shapes. After all, their work was pioneering, so no best-practice reports or guidelines were available.

The mode of resource utilization by each of the three initiatives is largely specific to the locality and history of the respective communities. Therefore, presented case studies do not provide blueprints for initiatives in other localities. Nevertheless, common factors for success that were identified, such as the choice of a location-based approach, the use of progressive and integrative information campaigns, and flexibility in maintained activities, are hoped to support the successful implementation of initiatives in other communities.

References

Bavarian State Office for Statistics. (2018). Fläche Gemeinden 2015.01.01 (community areas 2015.01.01), Genesis online database. Fürth/Munich, Germany: BSOS. Retrieved Jan 20, 2018, from https://www.statistikdaten.bayern.de/genesis/online?sequenz=tabelleErgebnis&selectionname=11111-001r.

Bund Naturschutz in Bayern e.V. (2004). Zahlen und Fakten: Berglandwirtschaft (numbers and facts: alpine farming). Munich, Germany: BNB e.V. Retrieved Jan 20, 2018, from https://www.bund-naturschutz.de/fileadmin/Bilder_und_Dokumente/Themen/Alpen/Zahle_Fakten_Berglandwirtschaft_Alpen.pdf.

City of Templin. (2017a). Integriertes Stadtentwicklungskonzept Templin 2030 (integrated city development concept Templin 2030). Templin, Germany: CT. Retrieved Dec 20, 2017, from http://templin.de/media/1/Stadtentwicklung/insek-templin-2030-endfassung.pdf.

City of Templin. (2017b). Internet presence of Templin's city administration—History. Retrieved Dec 20, 2017, from http://templin.de/wir-uber-uns/geschichte

Community Administrative Union Schönau im Schwarzwald (CAU Schönau). (2017). Jahresabschluss 2016—Kernhaushalt (annual accounts 2016—core household). Schönau im Schwarzwald, Germany: CAU Schönau. Retrieved Jan 20, 2018, from http://www.gvvschoenau.de/pb/Stadt_Schoenau,Lde/Stadt+Schoenau/rathaus+_+politik/Finanzen.html.

Community Administrative Union Schönau im Schwarzwald (CAU Schönau). (2018b). Internet presence of Schönau im Schwarzwald's city administration—History, Public facilities, Energy experience Schönau im Schwarzwald. Retrieved Jan 20, 2018, from http://www.gvvschoenau.de/pb/Stadt_Schoenau,Lde/Stadt+Schoenau.html

Community Administrative Union Schönau im Schwarzwald (CAU Schönau). (2018a). Internet presence of the Schwarzwaldregion Belchen - City profile of Schönau im Schwarzwald (http://www.schwarzwaldregion-belchen.de/de/orte/schoenau.php) Accessed 20 January 2018.

Community of Bad Hindelang. (1997). Tourismus Bericht 1996/1997 (tourism report 1996/1997). Bad Hindelang, Germany: CBH.

Community of Bad Hindelang. (2006). Hindelang Natur & Kultur—die Verbindung zwischen extensiver Berglandwirtschaft und dauerhaftem Tourismus (Hindelang Nature & Culture—the connection between extensive alpine farming and sustainable tourism). Bad Hindelang, Germany: CBH.

Community of Bad Hindelang. (2014). Tourismus Bericht 2013 (tourism report 2013). Bad Hindelang, Germany: CBH. Retrieved Jan 20, 2023, from http://marktgemeinde.badhindelang.de/buergerservice/tourismusberichte.html.

Community of Bad Hindelang. (2017). Tourismus Bericht 2016/2017 (tourism report 2016/2017). Bad Hindelang, Germany: CBH. Retrieved Jan 20, 2023, from http://marktgemeinde.badhindelang.de/buergerservice/tourismusberichte.html.

Community of Bad Hindelang. (2018a). Internet presence of Bad Hindelang's community administration—Key Fig., Economic development. Retrieved Jan 20, 2018, from http://marktgemeinde.badhindelang.de/buergerservice.html.

Community of Bad Hindelang. (2018b). Internet presence of Bad Hindelang's community administration—Town description. Retrieved Jan 20, 2018, from http://presse.badhindelang.de/allgemeine-informationen/ortsbeschreibungen/ortsbeschreibungen/article/oekomodell-hindelang-nachhaltige-berglandwirtschaft-garantiert-dauerhaften-tourismus-10632.html.

Community of Bad Hindelang. (Unpublished). Bewerbung zur Aufnahme in das Verzeichnis des immateriellen Kulturerbes in Deutschland (Application for the inclusion into the German Nationwide Inventory of Intangible Cultural Heritage). Bad Hindelang, Germany: CBH.

Department for Food, Agriculture and Forestry Bavaria. (Unpublished). Data sheet on government-sponsored farms and their area in Bad Hindelang in 2010, 2015, and 2017.

Elektrizitätswerke Schönau GmbH (EWS). (2017). Geschäftsbericht 2016 (business report 2016). Schönau im Schwarzwald, Germany: EWS. Retrieved Jan 20, 2018, from https://www.ews-schoenau.de/ews/genossenschaft/geschaeftsberichte/.

Elektrizitätswerke Schönau GmbH (EWS). (2018). Internet presence of the EWS—History, Subsidy program. Retrieved Jan 20, 2018, from https://www.ews-schoenau.de/ews/geschichte/.

Federal Agency for Nature Conservation. (2007). Regionalvermarktung in den deutschen Biosphärenreservaten—Status Quo-Analyse und Optimierungspotentiale (regional marketing in German biosphere reserves—status quo analysis and optimation potentials). Bonn, Germany: FANC.

Federal Bureau of Statistics. (2016). Datenreport 2016—Räumliche Mobilität und regionale Unterschiede (data report 2016—spatial mobility and regional disparities). Bonn, Germany: FBS.

Federal Institute for Research on Building, Urban Affairs and Spatial Development. (2017). Beschäftigte Primärer Sektor 2014—Ländlicher Raum (employees primary sector—rural area). Bonn, Germany: FIRBUASD.

Federal Ministry of Food and Agriculture. (2016b). Landwirtschaftliche Betriebe nach Größenklassen der landwirtschaftlichen Fläche, in *Statistisches Jahrbuch über Ernährung, Landwirtschaft und Forsten* C, I. Bonn, Germany: FMFA.

Federal Ministry of Food and Agriculture. (2016a). Landwirtschaft verstehen—Fakten und Hintergründe (understanding agriculture—facts and backgrounds). Berlin, Germany: FMFA. Retrieved Jan 20, 2018, from https://www.bmel.de/SharedDocs/Downloads/Broschueren/Landwirtschaft-verstehen.pdf?__blob=publicationFile.

Graichen, P. (2003). *Kommunale Energiepolitik und die Umweltbewegung—Eine Public-Choice-Analyse der "Stromrebellen" von Schönau (communal energy politics and the environmental movement—a public choice analysis of the "power rebels" of Schönau).* Campus Verlag.

Hofmann, G., & Pommer, U. (2013). Die Waldvegetation Nordostdeutschlands (the forest vegetation of northern Germany), in *Eberswalder Forstliche Schriftenreihe* 54. Potsdam, Germany: Ministerium für Infrastruktur und Landwirtschaft des Landes Brandenburg, Landesbetrieb Forst Brandenburg, Landeskompetenzzentrum Forst Eberswalde.

Horn, F. (2013). Almabtrieb im Allgäu—Runter mit dem Rind (bringing down cattle from alpine patures). *Spiegel Online*. Retrieved Jan 20, 2018, from http://www.spiegel.de/reise/deutschland/allgaeu-almabtrieb-als-festtag-a-921376.html.

Hupfer, M., & Nixdorf B. (2011). Zustand und Entwicklung von Seen in Berlin und Brandenburg (status and development of lakes in Berlin and Brandenburg). Berlin, Germany: Berlin-Brandenburgische Akademie der Wissenschaften, Berlin.

German Commission for UNESCO. (2017). Internet presence of the German Commission for UNESCO—Allgäu's high alpine agriculture in Bad Hindelang. Retrieved Dec 20, 2017, from http://www.unesco.de/en/kultur/immaterielles-kulturerbe/german-inventory/best-practice-register/allgaeus-high-alpine-agriculture-in-bad-hindelang.html.

Janzing, B., Seifried D., Rasper M. (2008). Störfall mit Charme—Die Schönauer Stromrebellen im Widerstand gegen die Atomkraft (disruptive incidence with charme—the power rebels of Schönau in resistance to nuclear energy). Vöhrenbach, Germany: doldverlag.

Jeutner, M. (2016). Internet presence of the Stephanus Foundation. Retrieved Jan 20, 2018, from http://www.stephanus.org/aktuelles/rundschau/juni-2016/startseite/artikel/detail/auszeichnung-fuer-schulwaldfoerster-joachim-lange-in-templin-1.

Linder, K. (2000). Nachhaltige Gemeindeentwicklung am Beispiel der Allgäuer Berggemeinde Hindelang. *Mitteilungen der Fränkischen Geographischen Gesellschaft* 47. Erlangen, Germany, pp. 1–136.

Makowitz, B., Knitter, E., & Kunze, M. (2014). Templin—Eine Maerkische Stadt im Wandel der Geschichte (Templin – a city of the Mark through the course of history). Templin, Germany: City of Templin.

Manig, J. & Verch, C. (2004). Die Entwicklung des Kommunalwaldes seit 1945 und als Bestandteil des Volkswaldes: vom Bodenfonds bis zur Restitution (the development of community forest since 1945 and as a part of the people's forest: from land funds to the restitution), in Ministry for Agriculture, Nature Conservation, and Regional Planning Brandenburg, State Institute of Forestry Eberswalde (eds.), *Eberswalder Forstliche Schriftenreihe* XX. Berlin, Germany: hendrik bäßler verlag, pp. 15–24.

Naturpark Südschwarzwald e.V. (2018). Internet presence of Naturpark Südschwarzwald e.V.—Geology. Retrieved Jan 20, 2018, from http://www.naturpark-suedschwarzwald.de/natur/geologie.

Naturschutzbund Deutschland e.V. (2013). Zukunft gestalten im Kommunalwald—Neue Ideen verbinden Forstwirtschaft und Waldnaturschutz (shape the future in community forests—new ideas conncect forestry and forest conservation). Berlin, Germany: ND e.V.

Radkau, J. (2011). Eine kurze Geschichte der deutschen Antiatomkraftbewegung (a short history of the German anti-nuclear movement), in Federal Agency for Civic Education (ed.): *Aus Politik und Zeitgeschichte* 46–47. Bonn, Germany: Federal Agency for Civic Education.

Ringler, A., et al. (2014). Bad Hindelang: Deutschlandweit die alpflächenreichste Kommune (Bad Hindelang: Nationwide the community richest in alpine pastures). In W. Kleiner (Ed.), *Kulturerbe Alpwirtschaft in Bad Hindelang im Naturschutzgebiet Allgäuer Hochalpen* (pp. 52–63). Eberl Medien GmbH & Co.KG.

Schiffer, H-W. (1999). Energiemarkt Deutschland (energy market Germany). Cologne, Germany: TÜV Media GmbH.

Schmid, M. (Unpublished) Bachelor's thesis. Entwicklung und Bewertung des " Ökomodells Hindelang" anhand ausgewählter Indikatoren (development and evaluation of the "eco model Hindelang" with selected indicators). Institute for Geography, Friedrich-Alexander University Erlangen-Nürnberg, Germany.

State Office for Statistics Baden-Württemberg. (2018). Bevölkerung im Überblick - Gemeindegebiet, Bevölkerung und Bevölkerungsdichte seit 1961 (population at a glance—community area, population and population density since 1961)—Schönau im Schwarzwald (Kreis Lörrach). Stuttgart, Germany: SOSBW.

Waldhofschule Templin. (2018). Internet presence of the Waldhofschule Templin—Our profile, The forest project. Retrieved Jan 20, 2018, from http://www.waldhofschule.de.

Wirthensohn, E. (1988). Das Öko-Modell-Allgäu, dargestellt an der Gemeinde Hindelang (the eco model Allgäu, depicted through the community Hindelang). Ein Zwischenbericht. Buchenberg, Germany.

Advancing Sustainability in Agriculture

Soil Fertility of Paddy Fields in Japan and Organic Rice Cultivation Using SOFIX Analysis

Quoc Thinh Tran and Motoki Kubo

Abstract In Japan, rice should be cultivated through sustainable agricultural systems that reduce the use of chemical fertilisers and pesticides. This study investigates the soil fertility of paddy fields under various management practices to evaluate soil fertility in organic farming fields. These results indicate that paddy fields have a higher bacterial biomass and a narrower range of total carbon than orchards and upland fields. Variations in soil fertility can be explained by the similar management practices in paddy fields in Japan. The application of organic fertilisers improved the soil bacterial biomass and total carbon in paddy soils. Moreover, the nutrient contents, including total carbon, total phosphorous, and total potassium, were within suitable ranges for paddy soil fertility. Therefore, instead of using chemical fertilisers, cultivation systems can use organic fertilisers for sustainable development.

Keywords Bacterial biomass · Paddy field · Organic fertiliser · Total carbon · Soil fertility

1 Introduction

Over 90% of global rice production can be attributed to Asia-Pacific countries, such as China, India, Indonesia, Bangladesh, and Vietnam. In Japan, rice production accounts for approximately 50% of the agricultural area and represents a stable market. In addition, rice is a staple food, the sustainability of rice production is important for food safety.

Rice production in Japan strongly depends on chemical fertilisers and pesticides, the use of which has gradually increased after they replaced organic fertilisers in rice cultivation post-World War II (Katayama et al., 2015). Japan imports the raw

Q. T. Tran (✉) · M. Kubo
Department of Biotechnology, Faculty of Life Science, Ritsumeikan University, Shiga, Japan
e-mail: thinh-89@fc.ritsumei.ac.jp

M. Kubo
e-mail: kubo@sk.ritsumei.ac.jp

M. Yoshizumi et al. (eds.), *Navigating Local Sustainability in Food, Community, and Innovation*, https://doi.org/10.1007/978-981-96-8963-7_11

materials of chemical fertilisers for its agricultural practices. However, the price of chemical fertilisers has increased worldwide owing to the Russia-Ukraine War (Hassen & Bilali, 2022). Compared to chemical fertilisers, organic fertilisers can serve as a more environmentally and economically sustainable alternative for rice cultivation. In 1999, the Japanese government initiated a "law for promoting the introduction of sustainable agricultural production practices." More recently, the MIDORI program was established to support the development of organic farming in Japan. Given farmers' familiarity with the use of chemical fertilisers, the implementation of organic farming, especially the application of organic fertilisers, is difficult because of application rates or the selection of suitable organic fertilisers. In addition, inappropriate organic management practices might lead to substandard results in soil fertility and rice yield.

Previous studies (Parveen et al., 2004; Yamazaki et al., 2001) have investigated the negative effects of chemical fertilisers and pesticides on the surrounding environmental biodiversity. For instance, the long-term application of such chemicals is likely to degrade soil fertility. Soil fertility index (SOFIX) analysis techniques have been developed to evaluate soil fertility based on biological, chemical, and physical properties (Adhikari et al., 2014; Aoshima et al., 2006). SOFIX analysis has been used to evaluate soil fertility and develop organic cultivation in Japan (Islam et al., 2023; Pholkaw et al., 2019; Tran et al., 2021). This study used the SOFIX database to determine the soil fertility features in paddy fields. Further, SOFIX analysis was employed to evaluate soil fertility in organic rice cultivation.

2 Materials and Methods

The soil fertility database from our previous studies (Araki et al., 2016; Pholkaw et al., 2020) was obtained. A total of 117 paddy soil samples were collected in Japan from 2014 to 2018. Soil fertility was analysed using the SOFIX technique, and a database of soil fertility in upland and orchard fields was obtained for comparison.

The field experiment was conducted in Kusatsu City, Shiga Prefecture, Japan. Organic fertilisers (cow manure, horse manure, and rice bran) were applied in four fields with areas from 2,000 to 4,000 m^2 from 2019 to 2022. The soil samples were collected from October to November after harvesting, then analysed using the SOFIX technique. Bacterial biomass was estimated by quantifying environmental DNA (eDNA) using a slow stirring method (Aoshima et al., 2006). Two parameters of microbial activities (nitrogen (N) circulation and phosphorus (P) circulation activities) were examined according to previous studies (Adhikari et al., 2014; Islam et al., 2023); N circulation was calculated based on the values of NH_4^+–N oxidation and NO_3^-–N oxidation rates and bacterial biomass, while P circulation was calculated from degradation of phytic acid in soil. Total nitrogen (TN) and total phosphorus (TP) in paddy soils were determined using a UV visible spectrophotometer (U-1900 Spectrophotometer; Hitachi, Tokyo, Japan) after Kjeldahl digestion with copper sulfate, concentrated acid sulfuric, and hydrogen peroxide at 420 °C. Kjedahl solution was also used for measuring total potassium (TK) using the Z2300 atomic absorption spectrophotometer (Hitachi, Tokyo, Japan). Total carbon (TC) was analysed using the TOC Analyzer (Model: SSM-5000A, Shimadzu, Kyoto, Japan). To analyse mineral NO_3^-–N and NH_4^+–N, a soil-1 M KCl solution suspension (1:10, w/v) was shaken reciprocally at 100 rpm for 1 h and the extracts were analysed by the brucine and molybdenum blue methods, respectively. The available phosphorus was analysed by Truog method with 0.001 M H_2SO_4 (pH3). Exchangeable K was analysed by extraction with 1 M ammonium acetate (pH7). The pH and EC were determined in a 1:2.5 soil–water mixture (w/v). Soil pH was measured by a pH meter (LAQUA. F-71, Horiba, Kyoto, Japan). Soil EC was measured by an EC meter (5LE1-408, Kenis, Hyogo, Japan).

3 Results and Discussion

3.1 Soil Fertility in Paddy and Other Agricultural Fields

The TC and bacterial biomass in the paddy fields were not normally distributed (Fig. 1). Bacterial biomass varied widely, but the majority of the data were distributed within 0–20 $\times$ 10^8 cells/g soil. Bacterial biomass and TC in the upland and orchard fields were distributed over a wide range and showed similar tendencies, whereas the paddy fields presented specific features (Fig. 2). In particular, bacterial biomass was higher in paddy soils than in other fields. Bacterial biomass is strongly affected by agrochemical application and accumulation in the soil. In paddy fields, flooding periods may reduce the retention or accumulation of agrochemicals in the soil. In addition, the alternating flooding and wetting periods lead to a higher bacterial abundance in paddy soils. The constant return of rice straw as an organic matter after harvesting also increases bacterial biomass. In contrast, the range of TC content was narrow, mainly ranging from 10,000 to 20,000 mg/kg soil, with an average of 15,420 mg/kg. There was no correlation between bacterial biomass and TC in the paddy soils. This probably occurred because of the similar rice cultivation in Japan, where only one crop is cultivated per year. In addition, paddy fields might have been amended mainly with chemical more than organic fertilisers.

3.2 Suitable Soil Fertility Conditions According to SOFIX Analysis

Table 1 shows the average values of the soil fertility parameters (TC, TN, TP, TK, N circulation, and P circulation) for different levels of bacterial biomass. A bacterial biomass lower than 2 $\times$ 10^8 cells/g decreased the N and P circulation activities, especially the N circulation activities. In agreement with Kubo et al. (2014), low

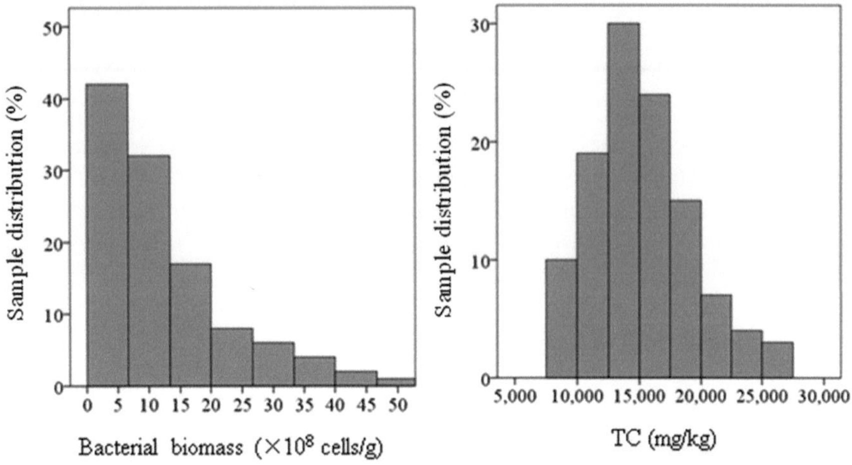

Fig. 1 Sample distribution according to bacterial biomass and TC in paddy fields

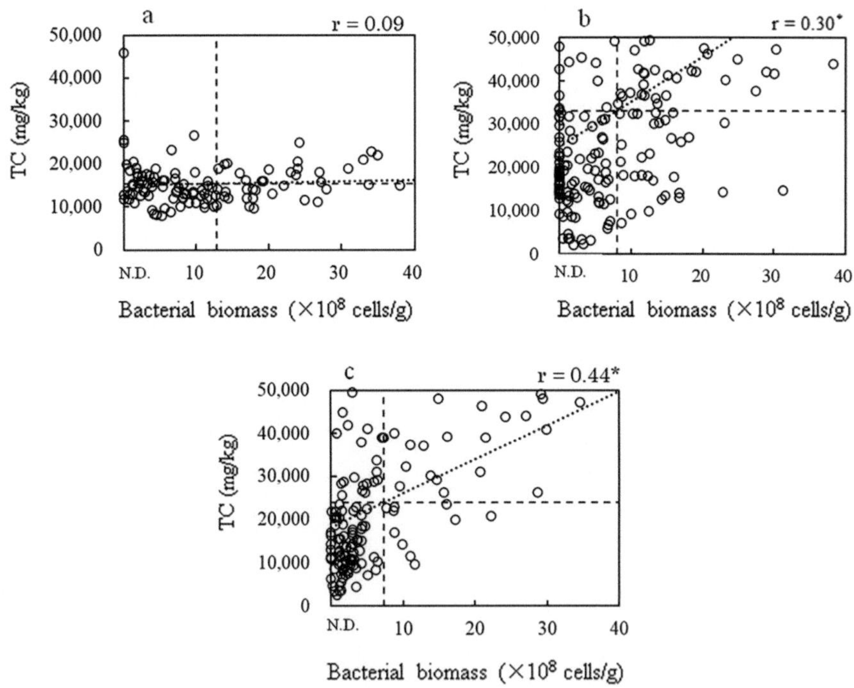

Fig. 2 Correlation between soil bacterial biomass and TC in paddy (**a**), upland (**b**), and orchard fields (**c**) (Araki et al., 2016; Pholkaw et al., 2020). Dashed lines indicate average values. N.D. = not detected (<6.6 × 10^8 cells/g soil)

Table 1 Variation of soil fertility in paddy field according to bacterial biomass

Bacterial biomass (× 10^8 cells/g)	TC (mg/kg)	TN (mg/kg)	TP (mg/kg)	TK (mg/kg)	N circulation (point)	P circulation (point)	Soil sample number
<2	18,220	950	640	2,670	3	18	17
2–4	14,760	990	930	2,950	10	41	14
4–6	13,420	870	890	3,840	17	32	11
>6	15,080	1,150	930	3,380	29	41	75

bacterial biomass inhibited nitrogen mineralisation in the soil. A bacterial biomass of 2–6 × 10^8 cells/g drastically increased the P circulation activity in paddy soils, but the N circulation activity remained low. Increases in N and P circulation activities were observed in paddy soils with a bacterial biomass higher than 6 × 10^8 cells/g. By contrast, there were no significant differences in TC, TN, TP, and TK at different levels of bacterial biomass. Therefore, the suitable conditions for soil fertility for bacterial biomass include a TC higher than 13,420 mg/kg, TN higher than 870 mg/kg, TP higher than 890 mg/kg, and TK higher than 2,950 mg/kg.

3.3 Case Study of Soil Fertility in Paddy Fields Treated with Organic Fertilisers

Table 2 shows the average soil fertility in four paddy fields in Kusatsu City, where organic cultivation was implemented from 2019 to 2022. At the beginning of the experiments (2019), the bacterial biomass was high, but N and P circulation activities were low. After three years of applying organic fertilisers, soil bacterial biomass dramatically increased and nutrient (N and P) circulation activities improved. The TC values also increased each year. In addition, TN, TP, TK, mineral nitrogen, available P, and exchangeable K contents were stable and higher than the suitable conditions listed in Table 2. The application of organic fertilisers slightly reduced the soil pH.

Figure 3 shows the distribution of bacterial biomass and TC data from 2019 to 2022. In comparison with the data from 2019, bacterial biomass and TC improved in the years following organic fertiliser treatments. Therefore, the results indicate that applying organic fertilisers can improve soil biological properties and maintain suitable nutrient values in paddy soils. That is, the SOFIX analysis revealed that using organic fertilisers can maintain stable rice yields in sustainable agricultural systems. The rice yields in the four fields were stable during organic cultivation treatments and were equal to those of conventional cultivation systems in the local area.

Table 2 Average soil fertility in paddy fields from 2019 to 2022

Parameter	2019	2020	2021	2022
Bacterial biomass ($\times 10^8$ cells/g)	9.5 ± 3.4	18.9 ± 7.0	9.2 ± 2.8	18.6 ± 3.6
N circulation (point)	15 ± 3	24 ± 2	19 ± 3	34 ± 4
P circulation (point)	16 ± 4	20 ± 7	18 ± 2	17 ± 5
TC (mg/kg)	15,960 ± 4,930	15,850 ± 2,800	18,730 ± 2,900	17,040 ± 2,020
TN (mg/kg)	1,380 ± 340	1,580 ± 290	1,650 ± 210	1,210 ± 440
TP (mg/kg)	1,250 ± 410	900 ± 130	1,070 ± 150	1,070 ± 300
TK (mg/kg)	6,830 ± 560	7,360 ± 570	7,750 ± 810	5,180 ± 1,270
NO_3^-–N (mg/kg)	0 ± 0	0 ± 0	0 ± 0	0 ± 0
NH_4^+–N (mg/kg)	1.0 ± 0.5	3.0 ± 3.0	0 ± 0	5.0 ± 3.7
Available P (mg/kg)	77 ± 13	68 ± 15	91 ± 31	84 ± 19
Exchangeable K (mg/kg)	200 ± 100	190 ± 30	250 ± 20	110 ± 63
pH	6.9 ± 0.2	7.2 ± 0.2	5.7 ± 0.4	5.8 ± 0.1
EC (dS/m)	0.07 ± 0.01	0.06 ± 0.03	0.13 ± 0.06	0.03 ± 0.02

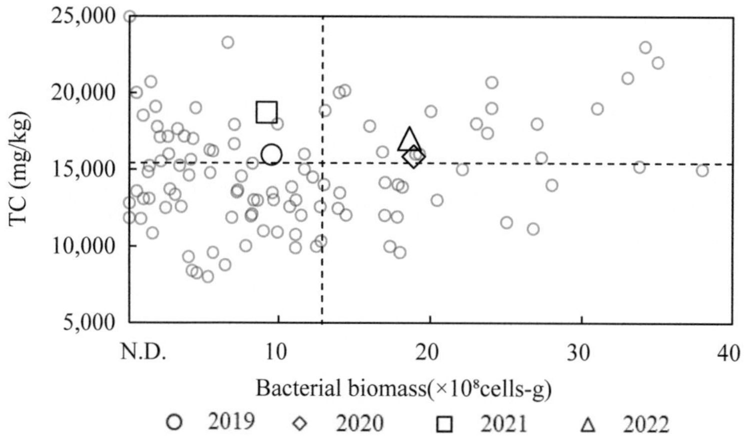

Fig. 3 Bacterial biomass and total carbon of soil samples from a paddy field in Kusatsu, Japan. Dashed lines indicate average values. N.D. = not detected (<6.6 × 10^8 cells/g soil)

4 Conclusions

This study clarifies and presents the features of soil fertility in paddy fields under organic cultivation based on a SOFIX analysis. Owing to monocropping and similar rice cultivation systems, soil fertility in paddy fields is different from that in other agricultural fields in Japan. Nevertheless, the short-term application of organic fertilisers improves soil fertility and is a promising solution for sustainable agricultural systems.

References

Adhikari, D., Kai, T., Mukai, M., Araki, K. S., & Kubo, M. (2014). Proposal for a new soil fertility index (SOFIX) for organic agriculture and development of a SOFIX database for agricultural fields. *Current Topics in Biotechnology, 8*, 81–91.

Aoshima, H., Kimura, A., Shibutani, A., Okada, C., Matsumiya, Y., & Kubo, M. (2006). Evaluation of soil bacterial biomass using environmental DNA extracted by slow-stirring method. *Applied Microbiology and Biotechnology, 71*, 875–880.

Araki, K., Perwira, I., Adhikari, D., & Kubo, M. (2016). Comparison of soil properties between upland and paddy fields based on the soil fertility index (SOFIX). *Trends in Microbiology, 8*, 81–91.

Hassen, T. B., & Bilali, H. E. (2022). Impacts of the Russia-Ukraine war on global food security: Towards more sustainable and resilient food systems? *Foods, 11*, 2301.

Islam, Z., Tran, Q. T., & Kubo, M. (2023). Development of a small-scale cherry tomato cultivation method using organic soil. *Organic Agriculture, 13*, 237–246.

Katayama, N., Baba, Y. G., Kusumoto, Y., & Tanaka, K. (2015). A review of post-war changes in rice farming and biodiversity in Japan. *Agricultural Systems, 132*, 73–84.

Kubo, M., Horii, S., Matsumoto, T., Mukai, M., & Adhikari, D. (2014). Evaluation of soil fertility for plant growth based on bacterial biomass and material cycle on soil environment. In J. N. Govil (Ed.), *Fertilizer technology II: Biofertilizer* (pp. 147–160). Studium Press LLC.

Parveen, S., Nakagoshi, N., Kohguchi, T. (2004). Trends in the use of agricultural pesticides and the environmental risk-reduction status in Japan. An evaluation of the last 15 years. *Outlook on Agriculture, 33*, 177–89.

Pholkaw, P., Muraji, A., Maeda, K., Kawagoe, T., Kubota, K., Sanpa, S., Tran, Q., & Kubo, M. (2019). Utilization of wood biomass for organic soil based on the soil fertility index (SOFIX). *Journal of Agricultural Chemistry and Environment., 8*, 224–236.

Pholkaw, P., Tran, Q. T., Kai, T., Kawagoe, T., Kubota, K., Araki, K. S., & Kubo, M. (2020). Characterization of orchard fields based on soil fertility index (SOFIX). *Journal of Agricultural Chemistry and Environment, 9*, 159–176.

Tran, Q. T., Araki, K. S., & Kubo, M. (2021). An investigation of upland soil fertility from different soil types. *Annals of Agricultural Sciences, 66*, 101–108.

Yamazaki, M., Hamada, Y., Ibuka, T., Momii, T., & Kimura, M. (2001). Seasonal variations in the community structure of aquatic organisms in a paddy field under a long-term fertilizer trial. *Soil Science and Plant Nutrition, 47*, 587–599.

Promotion Policies and Challenges in Sustainable Rice Agriculture in Shiga Prefecture, Japan

Thanh Tam Ho and Koji Shimada

Abstract Japan is targeting to reach complete carbon neutrality by 2050 across various sectors, including agriculture. In the ongoing era of climate change, promoting strategic policies that ensure food security while minimising the environmental impact seems vital. The global outlook for chemical fertilisers prices expects prices to remain higher in the foreseeable future. Therefore, it is important and necessary to promote sustainable forms of agriculture, such as organic agriculture that is environmentally friendly. Japan's Ministry of Agriculture, Forestry and Fisheries has been encouraging environmentally friendly agriculture since 1992. The Basic Law on Food, Agriculture, and Rural Areas, which includes organic farming, was enacted in 1999. Government policies have evolved to promote national self-sufficiency, food security, and sustainable development. This chapter is a brief review of sustainable rice farming and its promotional policies and challenges, focusing on the Shiga Prefecture, one of Japan's leading prefectures for sustainable agriculture.

Keywords Sustainable agriculture · Rice farming · Direct payment scheme · Challenges · Japan

1 Introduction

Agriculture contributes significantly towards greenhouse gas emissions, directly through on-farm emissions associated with production and indirectly through land use changes resulting from agricultural expansion. Nevertheless, agricultural practices hold great potential as a nature-based solution to mitigate climate change. Transitioning agriculture towards environmentally friendly and sustainable methods can

T. T. Ho (✉)
College of International Relations, Ritsumeikan University, Kyoto, Japan
e-mail: hotam@fc.ritsumei.ac.jp

K. Shimada
College of Economics, Ritsumeikan University, Kyoto, Japan
e-mail: shimada@ec.ritsumei.ac.jp

M. Yoshizumi et al. (eds.), *Navigating Local Sustainability in Food, Community, and Innovation*, https://doi.org/10.1007/978-981-96-8963-7_12

help farmers lower their on-farm carbon emissions, sequester atmospheric carbon dioxide in their soils, and enhance soil health.

Sustainable agriculture encompasses a variety of practices aimed at mitigating the environmental impact of farming and optimising the efficient use of input resources (USDA, 2022). These practices include minimising the use of chemical fertilisers, reducing reliance on chemically synthesised pesticides, and improving soil health. The Japanese Ministry of Agriculture, Forestry and Fisheries (MAFF) defines environmentally friendly agriculture (EFA) as sustainable agriculture that leverages the material circulation function of agriculture, focuses on harmony with productivity, and considers the reduction in environmental impact through reduced use of chemical fertilisers, pesticides, and soil preparation. Conversely, organic agriculture refers to farming methods that avoid the use of chemically synthesised fertilisers, pesticides, and genetic recombination technologies, and aim to reduce the environmental load as much as possible (MAFF, 2006). Japan has a century-long history of rice cultivation, which is a crucial part of its culture and economy. Although total methane (CH_4) emissions in 2020 were about 36% lower than those in 1990, CH_4 emissions from the rice sector accounted for approximately 40% of all methane emissions in Japan (MAFF, 2023). The use of sustainable agricultural practices (SAPs) contributes to environmental conservation and national neutrality goals and ensures long-term sustainability of agricultural systems. Japan first promoted EFA through the 1992 Direction of the New Policy for Food, Agriculture, and Rural Areas. In 1999, the Basic Law on Food, Agriculture, and Rural Areas was enacted, and it included organic farming. A system conforming to the 'Organic Japanese Agricultural Standard (JAS)' has been in operation since 1999, administered by a third-party organization in accordance with 'the Act on JAS'. Similar systems were implemented in Europe and the United States. The Act on Promotion of Organic Agriculture was enacted in 2006 to develop organic agriculture, clarify the responsibilities of national and local governments, lay the groundwork for measures to promote organic agriculture and foster its development. In March 2020, MAFF revised the Basic Plan for Agricultural Production and Management to boost organic agriculture in Japan. The revised Basic Plan's ambitious goal is to triple the number of organic farmers and land in the country by 2030. This is a major step aimed at promoting sustainable agriculture in Japan. Additionally, the Basic Plan proposes to increase the organic food market to JPY 328 billion by 2030, a significant increase from JPY 185 billion in 2017. Japan's local governments are encouraging organic agriculture and sustainable food systems, particularly in the wake of the COVID-19 pandemic.

Despite these efforts, Japan's contribution to organic farming in 2020 is only a small fraction (Organic World, 2022). As of 2020, only 11,992 ha of land were dedicated to organic farming from a total of 6,146,235 ha across Asia, constituting a mere 0.3% of the organic share of farmland. Likewise, during the same year, Japan had only 3,678 organic producers versus Asia's total of 1,809,628 producers. Hence, the need to accelerate organic agriculture in Japan to meet the goals set in the Basic Plan and foster sustainable agriculture practices in the country.

2 Sustainable Agriculture and Promotion Policies in Shiga Prefecture

In 2022, Shiga Prefecture was ranked 17th in total rice production (156.2 thousand tons) and 18th in total area under rice cultivation (47,100 ha). However, in terms of the ratio of paddy to rice, Shiga ranks second after Toyama Prefecture, with 93% of its total agricultural land dedicated to rice cultivation (46,900 ha). Additionally, Shiga has the largest area of environmentally friendly rice in Japan, at approximately 12,987 ha in 2022 (MAFF, 2022).

Lake Biwa located in Shiga Prefecture and known as the largest freshwater lake in Japan, serves as a vital water source for approximately 1.1 million people in Kyoto, Osaka, and other Keihanshin areas. It is deemed the 'water of life' for its ecosystem.

Recognising the importance of Lake Biwa and its biodiversity as a water source, the Shiga government has enacted several laws and policies to protect Lake Biwa and its biodiversity. These include the Ordinance for Eutrophication Prevention in Lake Biwa (1979), Special Law for Preserving Lake Water (1985), Lake Water Quality Conservation Plan (1987), Master Plan for Environmental Management of Yodo River System (1990), and Reed Colony Conservation Ordinance (1992). Shiga Prefecture has also implemented several promotional programs and projects to ensure environmentally harmonious agricultural production and provide consumers with safer and more reliable agricultural products. They established the EFA Certification System in 2001 and enacted the Ordinance for Promotion of EFA in 2003. These initiatives have played a crucial role in supporting sustainable agriculture practices, which involve reducing pesticide and chemical fertiliser use and incorporating conservation techniques.

Following the introduction of the Ordinance, Shiga became the first prefecture in the country to introduce the EFA Direct Payment System in 2004. This system has helped farmers adopt environmentally friendly agricultural practices and encouraged them to produce environmentally friendly agricultural products. Shiga Prefecture has been promoting the 'Fish Cradle Rice Paddies Project' since 2006. Additionally, Measures to Conserve and Improve Land, Water, and Environment (MCILWE) were incorporated into the advanced efforts of Shiga Prefecture in 2007. This further strengthened the promotion of EFA in the region. From 2011, direct payment subsidies for EFA, referred to as the 'Direct Payment System', were introduced nationally (see Table 1). This grant has significantly impacted farmers practising sustainable agriculture in Shiga Prefecture. It facilitates the production of sustainable agricultural products and encourages other farmers to embrace sustainable practices. More specifically, farmers will receive JPY 30,000–80,000 yen per ha for implementing SAPs. These practices include numerous environmentally-friendly farming methods, such as integrated pest management (IPM), the use of slow-release fertilisers, cover cropping, living mulch, organic farming, and rice-fish systems, etc.

The area covered by direct payment grants in Shiga Prefecture accounted for 20% of the nation's total, ranking first in Japan in 2017, followed by Hokkaido

Table 1 Promotion policies for sustainable agriculture in Shiga Prefecture

Year	Policies
2001	Shiga's Vision for Agriculture and Forestry, especially EFA and EFA Certification System
2003	Shiga Prefecture Ordinance Promoting EFA
2004	Shiga Agri-Environmental Direct Payment Scheme
2006	Fish Cradle Rice Paddies Project
2007	MCILWE (developed by MAFF)
2011	National Policy on Direct Payment Program (developed by MAFF)
2018	Shiga's Omi Rice Production and Distribution Vision
2021	Shiga Prefecture Basic Plan for Agriculture and Fisheries Based on the Government's Definition of Sustainable Agriculture

Source Author's collection; MAFF: Ministry of Agriculture, Forestry and Fisheries

(19%) (Shiga Prefecture, 2019). Nevertheless, the percentage of sustainable agriculture has declined recently across Japan, especially in Shiga Prefecture. In 2020, the agricultural area covered by direct payment grants in Shiga Prefecture decreased to 12,987 ha, ranking second in Japan (MAFF, 2022). The prefecture has been successful in expanding EFA practices, with approximately 44% of the planted area of paddy field rice in the prefecture now being cultivated as an EFA product (see Fig. 1), and it is likely to become the largest area for EFA in Japan in 2021 (see Fig. 2).

The prefecture has also made substantial efforts towards biodiversity conservation. Almost a hundred production organisations display the Environmentally Friendly Agricultural Product Certification Mark when shipping products, and the total number of processed products using eco-friendly agricultural products reached 76 in 2017. In 2016, Shiga Prefecture allocated approximately JPY 600 million to

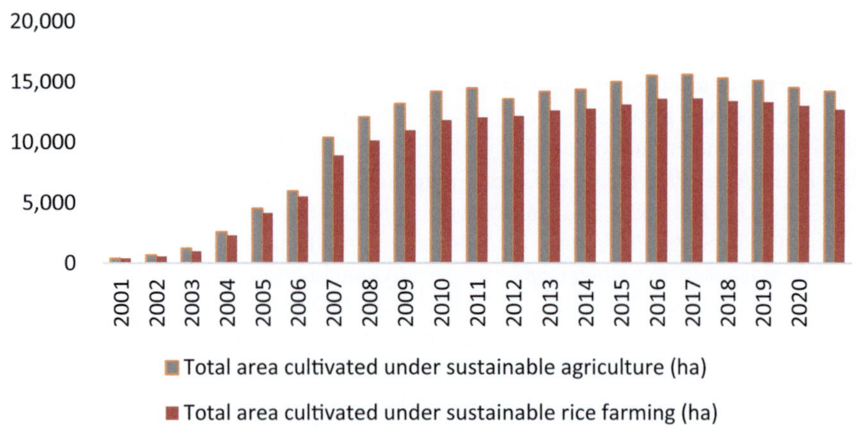

Fig. 1 Change in the total area cultivated under sustainable rice farming in Shiga Prefecture. *Source* Shiga Prefecture (2023b)

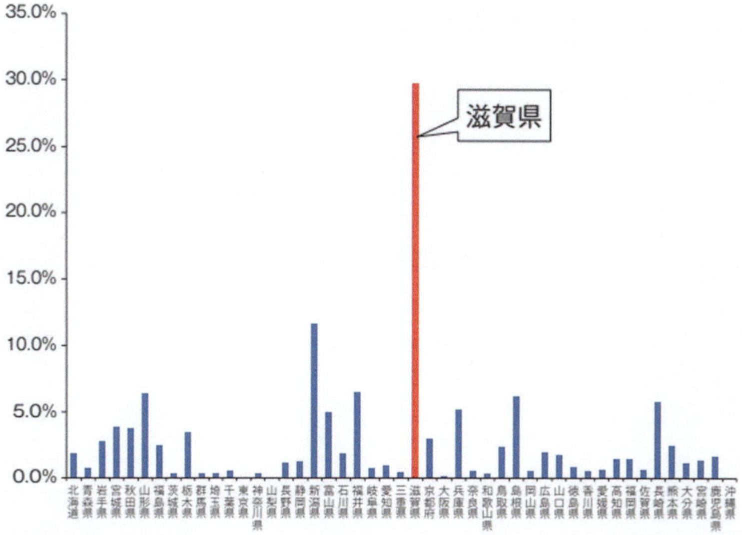

Fig. 2 Ratio of the area covered by the Direct Payment Subsidy for EFA to the Total Cultivated Land Area (2021). *Source* Shiga Prefecture (2023b)

a public conservation program for the ecosystem management of Lake Biwa, with the primary aim of mitigating the harmful effects of overgrown weeds and removing identified invasive species. The total subsidy for direct payment scheme in 2020 to 5,000 EFA farmers in Shiga Prefecture was around 55 million yen, with the national government contributing 50% (Shiga Prefecture, 2023a).To obtain certification for environmentally friendly rice, rice farmers in Shiga must meet four specific cultivation standards set by the central and local governments (see Fig. 3) (Shiga Prefecture, 2022). These standards ensure that rice is grown in an environmentally friendly manner.

1. The amount of chemically synthesised pesticides used must be less than half of the normal amount, and the total number of ingredients is seven or less. To reduce the use of chemically synthesised pesticides, Shiga rice farmers often use hot water disinfection of seeds to prevent seed borne diseases. Additionally, they do not use herbicides on the slopes of farm roads or drainage ditches adjacent to fields and around houses in management by mowers.
2. The amount of chemical fertiliser (nitrogen component) used is less than half of the normal amount, 4 kg/10a or less.
3. The environmentally friendly cultivation techniques adopted for Lake Biwa do not shed muddy water.
4. A record is maintained regarding the method used in rice cultivation.

Fig. 3 Paddy cultivated following EFA practices. *Source* Author

2.1 Fish Cradle Rice Paddies Program

Shiga Prefecture has been promoting the 'Fish Cradle Rice Paddies Project' since 2006, aiming to restore rice paddies so that both humans and fish coexist, recreating conditions that allow fish to move between Lake Biwa and the rice paddies for spawning and breeding (see Fig. 4). In 2015, the project was implemented in 27 regions across the prefecture covering approximately 127 ha. This can be considered an effective promotional program for environmentally friendly rice production and create a breeding ground for conserving lake fish such as the round crucian carp in Lake Biwa.

Furthermore, the 'Fish Cradle Rice Paddies' serve as an educational environment for children to learn about the ecosystem. Elementary school students had the opportunity to participate in juvenile-catching and wildlife observation events held in these paddies (see Fig. 5). To be entitled to use the trademark, rice farmers must meet the following conditions for fish cradle paddy rice and receive a certification (see Fig. 6a):

1. Only pesticides with the lowest toxicity levels to fish are used. After pesticide application, water valves are closed for few days to manage water flow and prevent chemicals from fleaving the fields, and fish from entering.
2. Rice fields are managed effectively to ensure fish habitats are preserved.

Fig. 4 Fish cradle paddy rice, namely Sakana-no-yurikago-suiden. *Source* Author

3. steps are taken to allow hatchlings to leave the paddies for the channels before the fields are drained.
4. Rice paddies host native fish species, which use the fishways established in the drainage canals for spawning.

2.2 An Overview of Omi Rice Production and Distribution

Based on the 'Omi Rice Production and Distribution Vision' formulated by the Omi Rice Promotion Association in March 2018, Shiga Prefecture adopted several measures to promote the cultivation and distribution of environmentally friendly rice, particularly the *Mizukagami* and *Koshihikari* rice varieties (see Fig. 6a).

Shiga Prefecture is actively promoting the expansion of Mizukagami production, in response to the growing demand for environmentally friendly rice. This is being done on the condition that the entire Mizukagami production is environmentally friendly. Consequently, in 2021, this variety of rice was cultivated over an expansive area of 3,254 ha in an eco-friendly agricultural product (Shiga Prefecture, 2023b). Additionally, Shiga Prefecture has focused on managing and promoting the distribution of Koshihikari rice separately from other rice varieties to ensure stable transactions with wholesalers and large-scale retailers, and this involves close cooperation with agricultural organisations and collection companies. To further promote environmentally friendly rice varieties, Shiga Prefecture are creating unique packaging for Koshihikari to market it as a new product. Furthermore, Shiga Prefecture disseminated information on the benefits of organic rice and fish cradle paddy rice,

Fig. 5 Educational programme for children on fish cradle rice paddies in Shiga Prefecture. *Source* Author

which are considered symbols of environment-friendly rice. Aiming to improve the image of environmentally friendly rice and increase its overall consumption, they are promoting various integrated distribution measures to create a brand of environmentally friendly rice, including 'Omi Rice' (see Fig. 5). For organic rice, Shiga Prefecture supports the introduction of riding paddy weeding machines and production technology support using the 'Organic Omi Rice Handbook' and sales efforts such as promoting the distribution and sales of 'Organic Omi Rice' using unified packaging (see Fig. 6b). As a result, the planted area is expected to expand to 269 ha (2021) (Shiga Prefecture, 2023b).

(a) (b) (c)

Fig. 6 **a** Fish cradle paddy rice (Koshihikari) produced in Shiga Prefecture, **b** Environmentally friendly rice (Mizukagami and Koshihikari), and **c** Organic rice (Mizukagami). *Source* Shiga Prefecture (2023a)

To promote the expansion, production, and distribution of sustainable agriculture products, priority promotion items have been determined for horticulture crops, such as vegetables. Integrated PR efforts are being implemented in all prefectures to expand distribution to markets and mass retailers and achieve superior sales. Several measures must be implemented to achieve the goals. First, the establishment of eco-focused agricultural product corners should be promoted at direct sales stores and mass retailers. This will help promote environmentally friendly agricultural products and increase their visibility. Additionally, the shipment of items produced in small quantities should be actively promoted to ensure that all environmentally friendly products are distributed. Furthermore, the lineup of environmentally friendly agricultural products should be strengthened to ensure a wide range of options are available to consumers. By promoting the production and distribution of environmentally friendly products such as fruits and vegetables, Shiga Prefecture can reduce the use of pesticides and chemical fertilisers and promote the use of SAPs. These efforts are crucial for promoting sustainable agriculture in Japan and ensuring that the agricultural sector continues to thrive in the face of climate change and other challenges. By investing in research and development and promoting education and raising awareness, Shiga Prefecture can build a more sustainable and resilient agricultural sector for the future. Producers are now actively displaying their certification marks for environmentally friendly agricultural products, thus, encouraging shipment and sale of the products. Businesses within and outside the prefecture, such as restaurants and business cafeterias, are also being encouraged to use environmentally friendly agricultural products. The 'Oishiga Ureshiga' campaign is promoting the sale and handling of environmentally friendly agricultural products by registered businesses. Furthermore, there is a specific focus on businesses in the watershed of Lake Biwa, emphasising the slogan 'Protect Lake Biwa by Eating (or Eat Eco Shiga)'. This initiative aims to raise awareness on the connection between consuming environmentally friendly agricultural products and preserving the water quality of Lake Biwa. These efforts, combined with consumer awareness and support, are contributing to the long-term sustainability of agriculture in the Shiga Prefecture and promoting a more environmentally friendly approach towards farming.

3 Benefits and Challenges of Sustainable Agriculture in Shiga Prefecture

3.1 Benefits of Sustainable Rice Agriculture

According to a report by MAFF (2023), sustainable agriculture decreases carbon emissions by approximately 143,393 tons of CO_2 equivalent annually. It is now more common to merge various SAPs with organic farming techniques. At the same time, it meets the demand for crop nutrients and enhances overall crop productivity. Four agricultural cooperatives (JA) in four cities in Shiga Prefecture—Takashima, Nagahama, Higashiomi, and Omihachiman—were selected for conducting questionnaire surveys with 112 rice farmers. Most of the surveyed farms (61%) had medium-scale rice fields (1–5 ha).

About 69% of surveyed rice farmers implemented SAPs for providing safe agricultural products (58%), followed by receiving direct payment subsidies (56%) and environment conservation, especially Lake Biwa (52%). Commonly applied SAPs in Shiga Prefecture include IPM, manual weeding of ridges (see Fig. 7), long-term drying (63%), slow-release fertilisers (53%), and organic farming (33%). Furthermore, most farmers (89%) implemented a combination of SAPS, while 11% implemented only one SAP.

Meanwhile, 31% of the surveyed rice farmers continue to cultivate conventional rice, citing reasons such as excessive efforts required for sustainable rice (45%) cultivation and insufficient direct payment subsidies for production cost compensation (36%).

Compared to 2000, Shiga Prefecture has seen a reduction in the use of chemically synthesised pesticides by approximately 40%. Furthermore, the load of total nitrogen originating from agricultural land and entering the Lake Biwa environment reduced by 18% in 2015 (Shiga Prefecture, 2019). According to Ho et al. (2023), implementing SAPs can improve soil fertility (i.e. bacterial biomass, total carbon (TC), and communities) compared to conventional farming practices in Shiga Prefecture. More specifically, the bacterial biomass in sustainable rice fields ranges from 14.5–23.3 (108 cell/g), which is higher than the 8.8–11.5 (108 cell/g) found in conventional rice fields. It is noteworthy that TC is significantly higher in sustainable rice fields with values ranging from 14,850 to 20,074 mg/kg while TC in conventional rice fields is lower, with values ranging from 13,560 to 19,190 mg/kg.

Fig. 7 Paddy field weeder for cultivating sustainable rice in Shiga Prefecture. *Source* Author

3.2 Challenges to Sustainable Agriculture in Shiga Prefecture

Shiga rice farmers are currently facing several challenges, including labour shortages, decline in rice demand, low consumption of environmentally friendly rice products, low profitability for environmentally friendly rice, higher prices of chemical fertilisers, and promotion policies. It is crucial to address these issues to support and sustain the agricultural sector in Shiga Prefecture.

3.2.1 Decline in Rural Population and Aging of Rural Communities

Population aging, low birthrates, and population decline are progressing rapidly in rural areas, particularly in the hilly and mountainous regions than urban areas. These demographic shifts pose significant challenges for the agricultural sector not only in Shiga but also across Japan. The depopulation of farming communities and labour shortages are major concerns that need to be addressed. In 2020, the number of farm household members in Shiga was 50,695, ranking 31st in Japan. On the national scale, Shiga ranks 43rd in terms of the number of key agricultural workers, with 9,961 individuals. Notably, a significant proportion of the existing agricultural workforce in Shiga is composed of individuals aged 65 and above. According to MAFF data (2022), approximately 78% of key agricultural workers in Shiga Prefecture belong to this age group. This demographic trend raises concerns about the future sustainability of the agricultural sector, as an ageing workforce may face challenges in meeting the demands of modern farming practices and adapting to new technologies. To overcome these challenges, it is crucial to implement strategies that revitalise rural areas, attract the younger generation to engage in agriculture activities, and provide adequate support and resources to farmers. Some farmers interviewed in Shiga Prefecture also suggested that the government should enact additional promotional policies to attract the younger generation into the agriculture sector (Ho, 2023).

3.2.2 Low Awareness of Environmentally Resilient Agricultural Products and Consumption

At the national level, rice consumption in Japan has been decreasing by over 100,000 tons (approximately 1.5%) annually in recent years (Kyodo News, 2022). In July 2022, Japan's rice demand for rice was estimated at 6.9 million tons, dropping below the threshold of 7 million tons for the first time. The COVID-19 pandemic has also altered dietary habits significantly. Consequently, demand for ready-to-eat meals and restaurant businesses has decreased.

Awareness regarding environmentally friendly agricultural products remains low among consumers, with only 18% of the surveyed consumers across Japan buying organic agricultural products (MAFF, 2019), and 45.7% respondents in the 2018 Shiga Prefectural Government Public Opinion Survey indicating their awareness about organic agricultural products. Additionally, only 27% consumers continue to purchase ecofriendly products. This highlights the need to promote consumer understanding and increase the demand for both organic and EFA products.

3.2.3 Low Profitability of Environmentally Friendly Rice

In Shiga Prefecture, environmentally friendly rice may lead to a slight increase in farmers' profitability by JPY 15,400/ha (approximately 102.4 USD[1]/ha) (see Table 2). This amount includes the direct payment subsidy and excludes the opportunity cost of family labour, when compared to conventional rice (Ho, 2023).

The cultivation of EFA such as *Mizukagami* is gaining momentum in Japan, although several challenges continue to limit its growth. One of the main obstacles is the instability in yield and quality, which can be less consistent compared to conventional cultivation. This issue, combined with increased costs and labour, can make it difficult for farmers to produce environmentally friendly rice. Another challenge is the small production volume of *Mizukagami* compared with the demand volume, leading to the need to expand production. Unfortunately, the direct payment

[1] Current exchange rate: 1 JPY = 0.0067 USD.

Table 2 A comparison of sustainable farming rice and conventional rice production

Item	Sustainable rice	Conventional rice
Average yield (kg/ha)	5,023.00	5,247.00
Selling price (JPY/60 kg)	10,700	10,400
Sale (JYP/ha)	891,670	866,670
Direct payment subsidy (JYP/ha)	40,000	0
Total income (JPY/ha)	931,670	866,670
Total cost (JPY/ha)	475,800	426,200
Profitability (JPY/ha)	455,870	440,470

Source Ho (2023)

system, which has played a significant role in the expansion of EFA efforts, currently has an insufficient budget because of the increase in nationwide efforts; however, the system is being reviewed.

3.2.4 The Challenges of Rising Prices of Chemical Fertilizers.

The global market has recently witnessed a significant surge in the prices of raw materials for chemical fertilisers. Since the end of 2021, prices have skyrocketed by 1.5–1.7 times, primarily driven by the increase in grain prices and the robust global demand for fertilizers (Statista Research Department, 2023). This sudden increase in fertiliser prices presents both challenges and opportunities for the agricultural sector.

Rising input costs are a challenge for farmers, as it has a direct impact on their profitability and cost-efficiency; however, this situation also opens avenues for exploring and adopting sustainable agriculture practices to mitigate the impact of increasing input costs. By reducing their reliance on chemical fertilisers and embracing eco-friendly alternatives, farmers safeguard themselves against rising prices and contribute towards environmental conservation. As such, the current scenario presents an opportune moment for farmers to reevaluate their practices and make conscious choices that align with sustainability principles.

Consistent with this, the findings from our 2022 questionnaire survey revealed that a substantial proportion of farmers in Shiga Prefecture are considering reducing the use of chemical fertilisers. Specifically, 42% of Shiga farmers expressed their intent to reduce their reliance on chemical fertilisers, while an additional 4% were committed to eliminate the use of chemical fertilisers. These data indicate a remarkable opportunity to promote and support sustainable agriculture in this region. By encouraging and facilitating the adoption of SAPs, stakeholders can collectively contribute towards environmental preservation, improve farmers' long-term profitability, and the overall development of a resilient and sustainable agricultural sector.

3.2.5 Challenges in Promotion Policies

The Acreage Reduction Policy

During the postwar era of food shortages in Japan, technological innovations during the 1940s aimed at addressing this critical issue. Technological advancements have played a crucial role in increasing agricultural productivity and alleviating food scarcity. However, as dietary habits changed during the mid-1960s, a surplus in rice production emerged, creating a new challenge for agriculture production. In 1971, the Japanese government implemented the Acreage Reduction Policy to combat overproduction and maintain stable rice prices. This policy aims to adjust rice production levels and prevent surplus production. The Acreage Reduction Policy, which has been in effect for five decades, has played a key role in regulating rice production. However, in 2018, recognising the changing dynamics and the needs of the agricultural sector, the government decided to abolish the policy, and this marked a shift towards a more market-oriented approach in rice production. Despite the abolition of the Acreage Reduction Policy, measures to prevent abrupt surges in rice production remain. Both local governments and agricultural cooperatives play an active role in guiding farmers, particularly in determining appropriate rice cultivation areas. By ensuring that the increase in production is gradual and sustainable, these measures aim to strike a balance between meeting rice demand and maintaining stability in the agriculture sector. The Japanese government's approach to rice production reflects its commitment to SAPs and the long-term viability of the sector. This approach not only supports farmers' economic well-being but also contributes to the overall stability and resilience of the agricultural sector.

Direct Payment System

Our 2022 questionnaire survey revealed that 48% of Shiga's rice farmers expressed their intent to pursue sustainable agriculture even in the absence of direct payment subsidies. However, 52% farmers stated that they would not implement sustainable agriculture. This suggests that, considering the habit of implementing SAPs among these farmers, the government could potentially reduce the amount of direct payment subsidies or reallocate them to alternative promotion policies. For instance, they could encourage transition to different rice varieties or other strategic crops. Confronted with a mix of challenges, including the low benefits of sustainable rice farming, a decline in consumption, and low selling prices, some farmers are gradually reducing their rice cultivation. Instead, they are starting to grow more profitable crops, such as wheat and soybeans. Other farmers are combining sustainable rice farming with conventional rice production (i.e. for staple food, rice processing, and whole-crop silage (WCS) for dairy farms). Direct payment schemes for EFA seem insufficient to support eco-farmers. Recently, policies have been revised to encourage the transition to other rice varieties and strategic crops. For example, rice farmers can receive JPY 35,000 per ha when they convert to growing wheat or soybeans, JPY 80,000 per ha for WCS feeding (livestock), JPY 20,000 per ha for processing rice, and JPY 55,000–105,000 per ha for rice flour.

4 Conclusion

Sustainable rice agriculture in Shiga Prefecture has broader environmental implications guidelines. Addressing these challenges requires multifaceted action from the local government, as discussed below.

(1) Investment in research and development to improve the yield and quality of environmentally friendly rice while reducing labour and production costs. Collaboration between government, industry, and academia can help promote EFA. For instance, industry partnerships with research institutions can result in new technologies that improve crop yield and quality.
(2) Attracting and retaining workers in the agricultural industry by providing training and educational opportunities, improving working conditions, and offering competitive wages. Additionally, promoting the benefits and rewards of working in agriculture can change the industry perception and attract more individuals, especially the younger generation, to join the workforce.
(3) Promoting the unique qualities and characteristics of Shiga rice, targeting new consumer segments, and exploring export opportunities to diversify and expand the market for rice products. In particular, collaborations with consumer-led organisations and businesses can help raise awareness and create the demand for locally grown rice. Additionally, efforts should be made to increase public awareness and education about the benefits of EFA. This can be achieved through various channels including social media campaigns, educational programs, and public events.
(4) Promoting smart agriculture. Technology has played a crucial role in promoting sustainable agriculture practices in Japan. For example, precision agriculture techniques, such as using drones and sensors, can help farmers monitor crop growth and identify potential issues early, as well as save labour. Additionally, government policies that incentivise environmentally friendly agricultural practices can encourage farmers to adopt such methods.

References

Ho, T. T. (2023). Sustainable Agricultural Practices: An empirical analysis in Shiga Prefecture, Japan. *Asia-Japan Research Academic Bulletin, 4.*

Ho, T. T., Tran, Q. T., & Shimada, K. (2023). Do sustainable agricultural practices improve rice productivity and soil fertility? An empirical study in Shiga Prefecture, Japan. *Chemical Engineering Transaction, 106*, 121–126.

Kyodo News. (2022). FOCUS: Japan Food Industry Eyes Rice Flour as Wheat Price Surges. Retrieved August 5, 2023, from https://english.kyodonews.net/news/2022/10/50f3126fd8a1-focus-japan-food-industry-eyes-rice-flour-as-wheat-prices-surge.html

Ministry of Agriculture, Forestry and Fisheries (MAFF). (2006). Act on the Promotion of Organic Agriculture (Article 2 in the Act No. 112 of 2006). Retrieved August 5, 2023, from https://www.maff.go.jp/j/seisan/kankyo/yuuki/attach/pdf/sesaku-1.pdf

Ministry of Agriculture, Forestry and Fisheries (MAFF). (2019). The Current State of Organic Agriculture in Japan in 2018 (in Japanese: 有機農業をめぐる事情). Retrieved August 5, 2023, from https://www.maff.go.jp/primaff/koho/seminar/2019/attach/pdf/190726_01.pdf

Ministry of Agriculture, Forestry and Fisheries (MAFF). (2022). Overview of Shiga Prefecture's Agriculture, Forestry, and Fisheries. Retrieved August 5, 2023, from https://www.maff.go.jp/j/kanbo/tiho/attach/pdf/todouhuken_gaiyou2022-25.pdf

Ministry of Agriculture, Forestry and Fisheries (MAFF). (2023). Japan's Strategy for Sustainable Food Systems (Strategy MIDORI), and Climate-Smart Agriculture. Retrieved August 5, 2023, from https://www.maff.go.jp/e/policies/env/env_policy/04_Strategy_MIDORI_and_Climate-Smart_Agriculture.pdf

Organic World. (2022). The World of Organic Agriculture: Statistics and Emerging Trends 2022. Retrieved August 5, 2023, from https://www.organic-world.net/yearbook/yearbook-2022.html

Shiga Prefecture. (2019). Environmentally Friendly Agriculture Promotion Basic Plan 2019. Retrieved August 5, 2023, from https://www.pref.shiga.lg.jp/file/attachment/5110999.pdf

Shiga Prefecture. (2022). Shiga's Organic Rice. Retrieved August 5, 2023, from https://shigaquo.jp/torikumi/environment/organic.html

Shiga Prefecture. (2023a) Environmentally Friendly Agricultural Products Certification System. Retrieved August 5, 2023, from https://www.pref.shiga.lg.jp/file/attachment/5386432.pdf

Shiga Prefecture. (2023b). Environmentally Friendly Agriculture Promotion Basic Plan 2023. Retrieved August 5, 2023, from https://www.pref.shiga.lg.jp/file/attachment/5404217.pdf

Statista Research Department. (2023). Monthly Fertilizer Price Development 2022, by Product. Retrieved August 5, 2023, from https://www.statista.com/statistics/1288602/quarterly-fertilizer-price-development-by-product/

United States Department of Agriculture (USDA), National Agricultural Library. (2022). Sustainable Agriculture. Retrieved August 5, 2023, from https://www.nal.usda.gov/farms-and-agricultural-production-systems/sustainable-agriculture#:~:text=Sustainable%20agriculture%20is%20farming%20in,best%20use%20of%20nonrenewable%20resources

Sustainable Agriculture for Small-Scale Farmers in Rural Thailand: The Practice of Farming for Self-sufficiency in Ban Na Isan

Ayako Fujieda and Siwaporn Chousorn

Abstract Agriculture is one of the major economic sectors in Thailand, with the export-oriented agriculture promoted by the government since the 1960s playing an important role in the country's development. While monoculture cash cropping has placed small-scale farmers at new risk, some farmers have been able to shift their farming style from cash cropping to self-sufficient farming by following His Majesty King Bhumibol Adulyadej's philosophy of 'Sufficiency Economy' and 'New Theory of Agriculture'. This chapter focuses on the practice of farming for self-sufficiency adopted in the village of Ban Na Isan, located at the east end of Chachoengsao Province in eastern Thailand, where shifting from monoculture cash cropping to farming for self-sufficiency has made the residents self-sustaining. A balanced use of farm land, crop diversification, and sustainable soil and water management has allowed farmers to be considerably self-sufficient on food and create cash income from selling extra crops. Therefore, farming for self-sufficiency has made both the life of farmers and the environment more sustainable.

Keywords Sustainable agriculture · Small-scale farmers · Self-sufficiency · Crop diversification · Thailand rural development

1 Background of the Study Area

The Kingdom of Thailand, with a population of approximately 60 million people, is a country in Southeast Asia that occupies the southern Indochinese Peninsula and the northern part of the Malay Peninsula. Under a government-led development plan

---

A. Fujieda (✉)
Faculty of Humanities, Kyoto Seika University, Kyoto, Japan
e-mail: afujieda@kyoto-seika.ac.jp

S. Chousorn
Kyoto University ASEAN Center, Kyoto, Japan
e-mail: chousorn.siwaporn.5w@kyoto-u.ac.jp

M. Yoshizumi et al. (eds.), *Navigating Local Sustainability in Food, Community, and Innovation*, https://doi.org/10.1007/978-981-96-8963-7_13

from the 1960s, the country has achieved steady economic growth despite experiencing economic depression and political instability. Currently, the annual per capita income in Thailand exceeds 5,640 USD, and the country is classified as an upper middle-income country. To attain the developed country status, the Thai Government announced its long-term economic goals using broad reforms in its 20-year National Strategy for 2017–2036.

Agriculture is one of the major sectors in Thailand, becoming the foundation of the country's economy. Rice is Thailand's most important agricultural product, being one of its major exporters in the world rice market. Other agricultural products produced in significant amounts include tapioca, rubber, grain, and sugar. Thailand is currently the 14th agricultural producer in the world, with a large agricultural land area of 211,000 km^2, which accounts for 43.3% of the total national land. According to the 2013 Agricultural Census, 5.9 million households, 25.2% of the total households in the country, engages in agriculture (National Statistical Office (1), 2013). Notably, 46.6% of the agricultural population is from the Isan region. The agricultural sector grew by an average of 4.1% per year between 1962 and 1983, with 70% of the population working in agriculture in 1980. Export-oriented agriculture has played an important role in the country's development, providing resources to support growth, supplying food to the urban workers, offering the basis for a dynamic and labour-intensive agro-processing industry, and fuelling the development of the rural non-food economy (Leturque & Wiggins, 2011).

Because of the country's climate and geography, the agricultural situation in Thailand is not uniform throughout its six regions (Fig. 1). The northern region shares a border with Laos and Myanmar and is geographically characterized by several mountain ranges and river valleys cutting through them. The communities in this region have developed in the basins surrounded by narrow mountain ranges. The highland areas are traditional habitats for hill tribes who practice shifting cultivation, while the lowland areas are relatively affluent areas of moderately fertile alluvial soil serviced by irrigation networks (Srimongkol & Marten, 1986). In the western region, high mountains and steep rivers are shared with Myanmar. It is less accessible and small-scale agriculture is practiced. The southern region occupying a part of the Malay Peninsula is mostly mountainous. Tin mining used to be a significant industry in this region, but it has been replaced by rubber plantation. The central region, composed of Chaophraya Delta, is one of the major rice farming areas in Thailand. The capital of Bangkok is located in this central region, and 20% of the total population lives in this metropolitan area. The northeastern region, commonly called Isan, is Thailand's largest region, located on the 55,000 km^2 Khorat Plateau. The population of Isan accounts for one third of the country's total population and half of the agricultural population. However, its agricultural productivity is relatively low due to the rainfall pattern, limited irrigation system, and poor soil condition in the region. Lastly, the eastern region is characterized by short mountain ranges alternating with small basins of short rivers that drain into the Gulf of Thailand. Fruit is a major component of the agriculture in this area.

Chachoengsao is an eastern province (changwat) that is 75 km to the east of Bangkok and covers an area of 5,351 km^2. The province has a history dating back

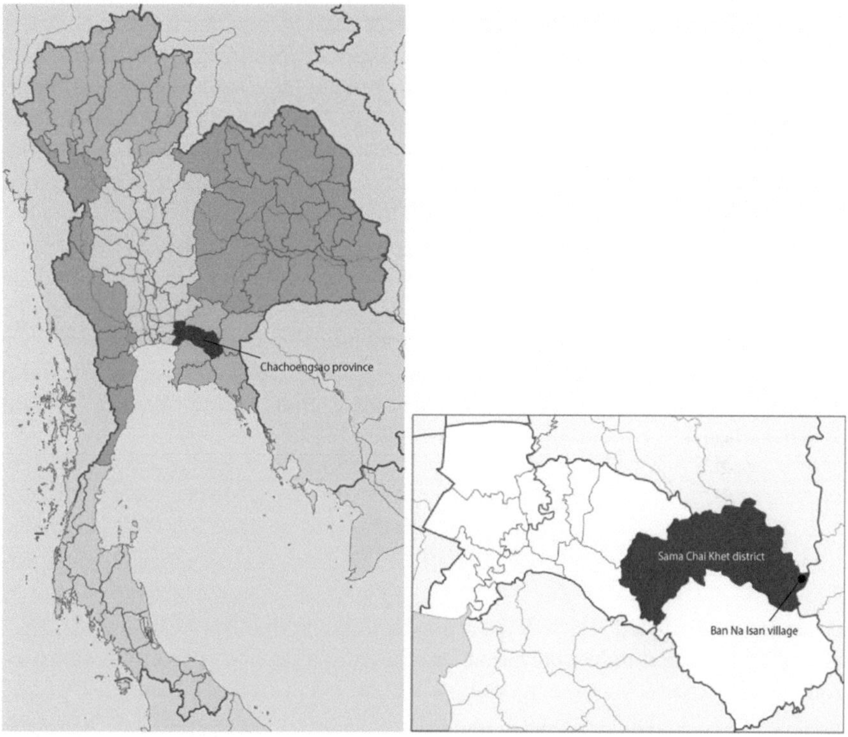

Fig. 1 Map of the six regions in Thailand and location of the Ban Na Isan village

to the reign of King Borommatrailokkanat in the Ayutthaya period, in which people originally settled by the Bang Pakong River. In the late 19th century, during a conflict between Siam and Annam over Cambodia, the Bang Pakong River was connected to the Chao Phraya River by a canal as a means of water transport for the army. Four canals branched off from the main canal to the north–south direction, resulting in the expansion of settlements and agricultural lands in the area (Pruksanubal, 2016). The western part of the province, belonging to the low river plain of the Bang Pakong River, is used extensively for rice farming, whereas the eastern part is hillier terrain. Sanam Chai Khet, a district (amphoe) in the eastern part of Chachoengsao, has four sub-districts (tambon) that are further subdivided into 70 administrative villages (muban). Sanam Chai Khet has a total population of 66,415, whom mostly engage in agriculture. The district's major agricultural products include rice, rubber, cassava, pineapple, mango, and eucalyptus (Chachoengsao, 2016).

Ban Na Isan is one of the villages in the Sama Chai Khet district of Chachoengsao Province. It shares a border with the Sa Kaeo Province. The village has about 0.8 km^2 of land and is situated next to the Khao Ang Rue Nai Wildlife Sanctuary, which covers over 1,000 km^2 of forest shared by the provinces of Chachoengsao, Chon Buri, Rayong, Chanthaburi, and Sa Kaeo. As of 2018, 226 households reside in

Ban Na Isan. The village is named after the fact that 90% of its residents migrated from provinces in the Isan region, namely Khon Kaen, Surin, and Buri Ram, during the 1970s. Although it is located in eastern Thailand, the Isan tradition and culture are observed in the village (Brahmasubha, 2012). The Ban Na Isan Border Patrol Police School is a primary school originally established by the village people that teaches primary education from grades one to six. Since it became under the control of the Border Patrol Police Bureau in 1994, the school has not only been providing education to school children but also serving as a source of knowledge for the local people and community (Brahmasubha, 2012). For secondary education, students from Ban Na Isan attend the Na Yao Secondary School in the neighbouring village. In this village, almost everyone engages in agriculture and grows cash crops such as cassava, corn, and eucalyptus. While most households depend on cash crop farming as a major source of income, twelve households in Ban Na Isan shifted to farming for self-sufficiency, continuing this practice for over a decade.

2 Challenges of Small-Scale Farmers

Agriculture contributed 36.3% of the gross domestic product in 1960. However, since the late 1980s, rapid industrialization has caused the decline of the gross domestic product and exports from the agriculture, forestry, and fisheries industries, despite nearly half of Thailand's labour population engaging in agriculture and the major industries in rural areas being agriculture and fishery. Subsequently, the share of agriculture in Thailand's gross domestic product (approximately 8–12%) has been more or less steady since the 1980s. Considering these circumstances, the Thai Government has targeted the promotion of agriculture, rural development, and food-related industries as one of the important elements in its development agenda.

Isan is the largest region in Thailand in terms of population and land area, with its 21.95 million people accounting for one-third of Thailand's total population in 2016. Approximately 70% of the population lives in rural areas, and this number continues to increase. Isan was recognized as a 'poverty area' in Thailand's first National Social and Economic Development Plan from 1961, and the government has intensively focused on the poverty reduction and socio-economic development of this region for over half a century. Consequently, absolute poverty in terms of nutrition, sanitation, and education has improved, but income inequality still remains. According to the 2013 Household Socioeconomic Survey, the average monthly household income in Isan was 19,181 Thai baht, the lowest among six regions, whereas the country's and Bangkok metropolitan area's average monthly household income were 25,194 and 43,058 Thai baht, respectively (National Statistical Office (2), 2013). Furthermore, despite 65% of the land for rice cultivation in Thailand being located in Isan, the net agriculture income per household in Isan is only one fifth of that in the central region. Currently, farmers in Thailand are divided into two types: large-scale farmers who produce products for export and sales to large agribusiness companies and small-scale farmers who produce products for the domestic market and private consumption.

Although small-scale farmers, with average land areas of 2.5–3.0 ha, account for 50% of the agricultural population, their production value is only 25% of the country's total agricultural production value.

Isan is the region where half of the farmers in Thailand live; however, their agricultural productivity is low and the life of farmers is difficult. Sandy soil derived from sandstone results in low fertility and water retention capacity. Salt accumulation in the soil due to elution from the underground rock salt layer is widely common and water resources are limited. The large-scale irrigation system in the region is underdeveloped due to the area's topography; thus, many farmers in Isan have limited access to water for agriculture, especially during the dry season. Therefore, majority of farmers in this area practice rainfall agriculture. However, this results in unstable crop yields because rainfall patterns differ annually. Recurrent droughts and floods are also challenges.

Cash crops such as maize cakes, eucalyptus, and rubber has been widely cultivated since the 1960s in the northeastern region. Practicing monoculture cropping may seem an easy method of gardening; however, the adverse effects of monocropping can lead to many problems down the road. Crop yields are good in the first few years, but then decrease yearly due to soil degradation and topsoil runoff as well as becoming afflicted by pests. Furthermore, the cost of agricultural materials, chemical fertilizers, and pesticides also increase. Moreover, the prices of export crops is often influenced by the demand of the importing country and situation of the global market. Although cash crop cultivation is believed to be an income generation opportunity, it actually does not guarantee a stable income. In the case of small-scale farmers, loans are taken from brokers to purchase their agricultural necessities or overcome short term financial difficulties with land as collateral. This leads to a large debt towards the brokers and creates financial hardship. Eventually, many small-scale farmers are forced to give up their land and quit farming.

Liam Buthehantha, a farmer from Ban Na Isan who has been practicing self-sufficient farming for almost twenty years, is an example of someone who had experienced hardship because of monoculture cash crop farming. Liam was born in 1956 as a son of a farm family in the Nong Ki district of Buri Ram Province. After working in Bangkok as a labourer for a couple of years in his teens, he returned home to start farming and cultivating cassava to sell for money and build a life with his family. However, cassava cultivation did not generate enough income, and it required him to continuously purchase chemical fertilizers and pesticides; this livelihood did not guarantee stable income because the price depended on the market. He had to borrow money from brokers to purchase food and necessities, and working diligently did not help him improve his family's living situation. By the time he was thirty years old, his only choices were to sell his farmland in Buri Ram or pass on.

In 1987, Liam decided to sell his land in Buri Ram to pay off his debts and moved with his wife and three year old son to Ban Na Isan in Chacheongsao Province, where government land was sold at a reasonable price. He purchased 50 rai (80,000 m^2; 1 rai = 1,600 m^2) of land in Ban Na Isan and started corn cultivation based on advice given by a local broker. However, earnings from corn cultivation were not enough to cover all expenditures for farming, which included equipment, fertilizer, pesticide, and daily necessities. His debts began to accumulate, and his alcohol and gambling habits returned as an escape from the financial hardship. Migration from Buri Ram to Ban Na Isan was supposed to be a fresh start to a new life with his family, but he fell into another debt trap. Recalling his previous situation, he stated that he did not possess the necessary knowledge to take control of his life and farming. Liam's experience was not exceptional. Numerous small-scale farmers in rural areas have practiced monoculture cash cropping and experienced similar hardships.

3 Introduction of a Community-Based Initiative

Sustainable agriculture for small-scale farmers in Thailand simply implies an agricultural system that will allow farmers to become self-sufficient, and can maintain a favourable ecological balance and stable farm communities (Jitsanguan, 2001). According to the interview with Liam from Ban Na Isan, His Majesty King Bhumibol Adulyadej's philosophy of 'Sufficiency Economy' and 'New Theory of Agriculture' are what they have adapted. How had twelve farmers in Ba Na Isan shifted their manner of farming from cash cropping to self-sufficient farming? Meeting with Wiboon Khemchalerm, a local pioneer of agroforestry, brought a change in their farming style.

3.1 Sufficiency Economy and the New Theory of Agriculture

His Majesty King Bhumibol Adulyadej had stressed the importance of economic self-reliance, based on the Buddhist belief in the middle path, in alleviating poverty in Thailand since 1974. While agricultural growth was one of the major development objectives, until the sixth National Economic and Social Development Plan, the agricultural policies of Thailand were intended to increase the production of rice for export and expand the area for commercial crops (Kasem & Thapa, 2012). However, this export-oriented economy that relies on foreign capital and external markets created an imbalanced system, causing unstable growth without self-sufficiency in Thailand. During the economic crisis in 1996–1997, His Majesty King Bhumibol Adulyadej reiterated and expanded on the concept of a sufficient economy. This philosophy guides people to live their lives according to the middle path and follows three core principles: moderation, reasonableness, and self-immunity. These principles are considered to be interconnected and interdependent (Mongsawad, 2010). In addition to these three principles, it highlights knowledge and morality as two key conditions to enable the sufficient economy. These concepts can be applied to the individual, community, and national levels. This philosophy highlights a balanced manner of living and guides the method to meet the challenges arising from globalization and changes and achieve a more resilient and sustainable economy (Piboolsravut,

2004). The following is a working definition of the philosophy (Wibulswasdi et al., 2012):

> Sufficiency Economy is a philosophy that stresses the middle path as an overriding principle for appropriate conduct by the populace at all levels. This applies to conduct starting from the level of families to communities and to the nation in terms of development and administration, so as to modernize in line with the forces of globalization. "Sufficiency" means moderation, reasonableness, and the need for self-immunity to protect from impacts arising from internal and external change. To achieve sufficiency, an application of knowledge with due consideration and prudence is essential. In particular, great care is needed in the utilization of theories and methodologies for planning and implementation in every step. At the same time, it is essential to strengthen the moral fiber of the nation, so that everyone, particularly public officials, academics, and business people at all levels, adhere first and foremost to the principles of honesty and integrity. In addition, a way of life based on patience, perseverance, diligence, wisdom and prudence is indispensable in creating balance and in coping appropriately with critical challenges arising from extensive and rapid socioeconomic, environmental, and cultural changes in the world.

In the agricultural sector, small-scale farmers who conventionally practice monoculture cash cropping, which depends on market prices, are vulnerable to external price shocks. These farmers often tend to overuse chemical fertilizer and pesticides to increase their production, consequently harming the environment. The New Theory of Agriculture is a practical example of applying the Sufficiency Economy philosophy, serving as a systematic guideline aimed to make farmers more self-reliant and self-sufficient using a strategy of holistic land management (Piboolsravut, 2004).

The theory recommends a three-stage procedure (Piboolsravut, 2004; Wibulswasdi et al., 2012). The first stage involves becoming self-reliant and self-sufficient at the household level by promoting a balanced use of farmland, crop diversification, and sustainable soil and water management. This allows farmers to be at least self-sufficient on food and generate cash income from selling extra crops. The theory proposes that 15 rai of land (24,000 m^2), the average land area of a household with four members, is to be divided into four parts at a ratio of 30:30:30:10. The first 30% should be converted to a 4 meter-deep pond, large enough to supply water even during the dry season. The pond can be used to cultivate fish and aquatic plants, which can be eaten at home or be sold. The next 30% is for cultivating rice capable of feeding one household. The other 30% should be used for growing vegetables, fruits, and trees for both subsistent and commercial purposes. The remaining 10% is for housing and other activities, such as livestock farming. Although the size and purpose have been carefully calculated, it can be adjusted to the local conditions. The second stage aims to strengthen the community by promoting cooperative activities such as establishing a community saving bank or community enterprises. This stage does not involve specific activities, allowing each community to adapt different activities depending on their situation. Such cooperative activities enhance the capacity of community members and allow them to become self-reliant by supporting the expansion of their production processes and commercial activities. The third stage encourages the community to expand their activities by connecting with outside sources. This includes the establishment of cooperative stores and rice-milling. It also encourages commercial banks and large enterprises with corporate social responsibility programs to support the community-based activities.

3.2 Pioneer of Sustainable Farming Practice

A local pioneer of integrated organic farming, Wiboon Khemchalerm (1963–2016) from Chacheongsao, took charge in changing the agricultural practice in Ba Na Isan. Once a cash crop farmer, he developed his own concept of sustainable farming, 'Wanakaset', which directly translates to 'forest agriculture' or 'agroforestry' (Green Net). His life-long work on sustainable farming has gained national recognition. He was selected as one of the members to prepare the national development plan in 1995, and served as a senator from 1996 to 2000 and a member of National Assembly in 2006.

Wiboon was born as the son of a farmer in 1963 and grew up in Koh Khanun, Chacheongsao. After working in Bangkok for several years, he returned home to begin cash crop farming when the Thai government adapted the first National Economic and Social Development Plan in 1961, which emphasized the development of agriculture and encouraged farmers to grow crops for export. Wiboon saw this new development plan as an opportunity for income generation and started cassava and sugarcane cultivation with chemical fertilizer and other technologies. In addition, he also became a broker to handle agricultural materials and chemicals and purchase agricultural products from the local farmers. Initially, he became successful and his earnings enabled him to expand his farmland, and he was even selected as a village leader of Ban Huai Hin. Eventually, he began to notice that his land did not produce as much as before and had to increase the amount of chemical fertilizers and pesticides. Although the inputs required for the cultivation increased, the price of crops remained almost the same. He began to borrow money, getting deeper in debt, and ultimately sold more than 300 rai (480,000 m^2) of land to pay off this debt. He turned his life around and focused on growing vegetables and plants such as morning glory for his own consumption in the remaining 9 rai (14,400 m^2) of land he possessed. Soon after, he witnessed that the humidity and fertility of his land gradually came back and realized that the land and forest was a source of food, medicine, and necessities for his daily living. From this simple beginning, he developed the Wanakaset concept, which focused on developing natural resources to meet one's own need and acquiring the necessary knowledge to use these resources efficiently (Green Net).

By focusing on existing biodiversity, the Wanakaset concept recommends growing seven layers of plants: (1) forest trees commonly found in Thailand, such as those in the Dipterocarpaceae family, (2) fruit-bearing trees such as mango, jackfruit, tamarind, or cinnamon, (3) shrubs, including santol, key lime, kaffir lime, banana, or papaya, (4) perennial plants such as lemon grass or mint, (5) aquatic plants, (6) root vegetables, such as turmeric or ginger, and (7) climbing plants. These layers can coexist, providing food, medicine, fuel, building materials, and household necessities while insuring the future (Green Net). According to a survey by the Food and Agriculture Organization of the United Nations (FAO), over 500 spices were identified in 9 rai of his land.

Based on his experience, Wiboon employed the practical principle of 'three knows, three plans, and five management' to enhance capacity of farmers (GotoKnow). 'Three knows' includes knowing yourself, knowing your problems, and knowing the resources to understand your life. It begins with understanding what one does, what the income source is, and what the expenditure is. The next step is knowing what one's problems are and what causes them. The last one is knowing what resources one has and the potential of one's knowledge and wisdom. Understanding yourself and having knowledge of your resources and potential can help build confidence. To solve problems and establish sustainable living, 'three plans' are recommended. These include a family plan as the foundation of life, a community plan for social stability, and a sustainable resource and environmental management plan. Furthermore, daily living requires five essential needs. To achieve self-reliance, 'five management' recommends people to be self-sufficient in five essential aspects: rice production, food, medicine, household goods, and soil fertility.

The Agroforestry for Social Foundation, a learning centre by Wiboon, was established to transfer his knowledge to local farmers and help them build their capacity by providing a demonstration forest, a library, a community trade centre as a learning platform, a coordination hub for the Eastern Forest Network, a knowledge centre for traditional medicine, and a camp school for children. Wiboon also travelled around and shared his experiences with famers' groups and NGOs throughout the country, including Ba Na Isan. A lecture by Wiboon at Ba Na Isan was the turning point of the farmers. Some of the farmers from Ba Na Isan settled at the Foundation and attended the seminars to learn about Wanakaset and obtain the necessary knowledge on agriculture and natural product production. Theis foundation has supported the farmers continuously and helped them to develop a self-sufficient manner of farming.

4 Initiatives and Outcomes

Currently, twelve households in Ba Na Isan engage in farming for self-sufficiency and self-reliance based on the New Theory of Agriculture and the Wanakaset concept. Liam is one of the pioneers who initiated this practice in Ba Na Isan in the late 1990s and has continued it for over twenty years. He currently lives with six family members, his wife, two sons, a daughter-in-law and two grandchildren, in a house called 'Baan Suan Oonson', which directly translates to 'happiness'.

4.1 Sustainable Agriculture Practice

From a monoculture cash crop farmer, Liam turned his life with debt around when he attended a three-day seminar by Wiboon in 1996. According to him, he was uninterested when he received an invitation to the seminar from a local school teacher. However, in the last day of the seminar, Wiboon stated that Thai farmers did not know themselves and told Liam to be more aware of his and his family's financial situation, considering his family's earnings and expenditures. Liam took this comment from Wiboon seriously and recorded his earnings and expenditures for one year. Evidently, their expenses were much more than their earnings. The record showed that the expenditures from Liam's family in one year was approximately 100,000 Thai baht, which included 60,000 Thai baht from his alcohol and gambling, 30,000 Thai baht for necessities, and 5,000 Thai baht for school supplies for the children. The earnings from corn cropping was 650 Thai baht per rai; when multiplied to his 50 rai of land, this was barely enough to cover the expenditure for necessities. It was an eye-opening experience for him and served as the catalyst to change his way of life. He decided to stop spending on alcohol and gambling, minimize his expenses, and grow rice and vegetables for personal consumption. He sold a majority of his land and kept 13 rai to grow rice, vegetables, and trees.

4.2 Land Use

Liam divided his land into two parts, 5 rai for rice and vegetable cultivation and 8 rai devoted to forest (Fig. 2). Originally a corn cropping farmland, this land was transformed into a forest with more than 300 species (Fig. 3 (1)). It has become an important source of food, medicine, and household goods. A pond located between the rice growing land and forest provides water for farming and is also used for fish cultivation (Fig. 3 (2)). The pond, which used to dry up during the dry season, is now filled with water throughout the year. In addition to growing rice in the field, a small area of land is allocated for vegetables and herbs (Fig. 3 (3)). The family has set up simple machines and equipment for rice processing, sugar processing, and oil extraction from plants and built a furnace for charcoal production (Fig. 3 (4)). These enable them to not only eat and use plants and vegetables directly, but also polish rice and process natural resources independently. The land the family owns in Ban Na Isan has expanded from 13 rai of the original land to 52 rai.

4.3 Essentials for Self-sufficiency

Liam followed the Wiboon's 'five management' philosophy by growing different plants to meet his five basic needs. These five essentials were rice as a staple food,

Fig. 2 Original land of Liam's family

Fig. 3 Areas in Liam's land: **a** forest (top left), **b** pond (top right), **c** rice and vegetable growing land (bottom left), **d** charcoal making kiln (bottom right)

other crops, medicine, household goods, and soil fertilizers. Riceberry, a kind of local brown rice, is cultivated during the rainy season; it is planted in June and harvested in November. The annual yield of riceberry is approximately 600 kg, which is more than enough for his family's consumption, and the surplus is packed and sold for 80 baht per bag. Liam's family is capable of seeding to harvesting and processing to selling by themselves. Apart from rice, a wide variety of vegetables, herbs, and edible plants grown in their garden and forest as well as fish from their pond are sources of food.

A rich dietary life is observed from the table. Figure 4 is one example of a lunch menu served during a visit to Baan Suan Oonson in January. It included two types of rice, riceberry and sticky rice, as a staple food, Thai style omelette with minced pork and leaves of ya nang (Tiliacora triandra), grilled tilapia, simmered snakehead with gourd, a chili-based condiment called nam phrik, tamarind paste, two kinds of banana, fruits from tala palm, and tea from sappanwood (Fig. 4). All the food except the eggs and pork meat were from their land.

A variety of medicinal plants are grown in the field and forest. Both Liam and his two sons possess profound knowledge on plants, and their efficacy is based on traditional Thai medicine as one of Liam's sons majored in alternative medicine in school and is an expert. Plant-based household goods such as kaffir lime shampoo, coconut soap, and toothpaste are all made at home. Organic fertilizer made out of bagasse, which residue from sugarcane refining, is also produced and used for farming.

Fig. 4 Lunch

(a) (b)

Fig. 5 **a** Home-made products made and sold by Liam's family, **b** list of products

4.4 *Income Generation*

The practice of farming for self-sufficiency has allowed Liam's family to acquire most of their food and necessities from their land. In addition to self-consumption, the family sells their extra crops and products. They have developed 18 kinds of plant-based products, which are shown in the list in Fig. 5. Wood vinegar, a byproduct from charcoal production or young plants, is also sold occasionally. These products are purchased by people who visit Ban Na Isan and are sold in organic products fairs and through the internet. The wide use of the internet and social media such as Facebook has made their products more accessible in recent years.

4.5 *Village Cooperatives*

In line with the second stage of three-stage procedure proposed in the New Theory of Agriculture by His Majesty King Bhumibol Adulyadej, community banks, rice banks (reserves), fish banks, and a community forest have been established in Ba Na Isan. In 1997, Liam and other community members decided to set up a community cooperative bank to financially support other community members by collecting money and offering loans with a low interest rate. This required the community members deposit a small amount of money to the community cooperative bank on the 15th day of every month. This initiative started with 32 households and expanded to all households in Ban Na Isan. However, when the government adapted the policy of 'village funding', which allocated one million Thai baht to each of Thailand's 78,000 villages as a means of providing working capital for locally-run rotating credit associations, the village decided to utilize it and the numbers of members participating in the community cooperative bank decreased. Currently, the community cooperative bank in Ban Na Isan is only operated at a small scale.

After the establishment of the community cooperative bank, a community rice bank was also established in Ba Na Isan in 1999. A problem of farmers who grow crops for commercial purpose was that they had to purchase rice for personal consumption. Occasionally, they have to sell their rice fields to earn cash and pay off debts. The rice bank was created to address this problem. The members deposit extra rice to the rice bank, which becomes available for the community members who do not have enough rice to eat. People who have received rice in previous years return rice with 20% interest during the Blessing Ceremony in February after the harvesting season. The extra rice collected as interest is sold and earnings from the sale are deposited to the community cooperative bank or offered to struggling households in the village. However, when the government set the minimum purchase price for rice, the members chose to sell rather than deposit rice. Consequently, the village rice bank is no longer available.

The community fish bank was another initiative made in Ban Na Isan. A ponded river was designated as a no-fishing area and fish were placed in this ponded river every 5th of December. It became a breeding area and plenty of fish flowing out of the area were available to catch, providing fish to the villagers all year around.

The community forest was another initiative aiming to maintain the forest around the village; it was based on the agricultural belief of the 'four mothers': mother of earth, mother of soil, mother of water, and mother of humans. Mature trees in the community forest were prohibited from being cut for sales, thereby allowing them to grow. Annual tree planting activities are conducted on Mother's Day in August and Father's Day in December. Every household with a new born baby sets out twenty plants and receives 500 Thai baht. In addition, the community forest in Ban Na Isan acts as a buffer zone for the nearby Khao Ang Rue Nai Wildlife Sanctuary, which is an elephant habitat. Elephants from the wildlife sanctuary become a threat to the local farmers because they sometimes invade farmlands and damage crops. Instead of preventing this occurrence using electrified barbed wire, poison, and firearms, the people from Ban Na Isan plant corns, sugarcane, and pineapples that the elephants like around the community forest and use it as a buffer zone. Furthermore, Ban Na Isan is a member of the eastern community forest network and shares their experience with other communities, educational institutions, and government institutions.

4.6 Knowledge Transfer

Initiatives by Liam's family and Ban Na Isan have become known outside of the village and have attracted visitors from outside after becoming the research subject of a university student from Thammasat University. Baan Suan Oonson, Liam's family home, is open to everyone and receives approximately 2,000 visitors annually, including children and adults, from the local schools, universities, enterprises, and others interested in learning about his initiatives and farming style. Visitors enjoy learning about their way of living, touring their forest and farm, experiencing rice planting, harvesting or processing, and making home-made products. Some visitors even stay from weeks to a few months as interns, learning about their farming knowledge, techniques, home production, and way of living. Liam's family has collaborated with a neighbouring family to host visitors who stay overnight in the village. This can be considered as another income opportunity, and they have been trying to develop a more systematic program to make the land a learning centre and an eco-tourism destination.

In addition to providing learning opportunities at their home, Liam is also a lecturer at the Mab Auang Agricultural Center, which was established in Chonburi by Wiwat Sailayamtorn. Wiwat is another person who has worked for more than 16 years for the Office of the Royal Development Project Board of the Prime Minister and served as the Deputy Minister of Agriculture and Cooperative. His centre offers an alternative model of education based on the philosophy of Sufficiency Economy and brings practical experience to farmers. The centre has seven objectives, having the roles of: (1) a research centre that develops technology for agriculture, aquaculture, livestock, energy, medicine, and environmental management based on local knowledge and wisdom; (2) a training centre to disseminate appropriate technology to farmers and government officials; (3) a technology transfer hub under the project of the Ministry of Agriculture and Cooperatives; (4) a demonstration site for the application of technologies; (5) a display for local agricultural equipment; (6) a tourist attraction for people interested in natural agriculture and; (7) a training centre open to government organizations and private enterprises. In addition to the Mab Auang Agricultural Center, Wiwat has established the Agri-Nature foundation, a network of over 50 agricultural research and learning centres all over Thailand. By being a lecturer at the Agricultural Center and keeping a close connection with other farmers and related organizations, Liam is able to share the knowledge and experience he obtained from his way of farming and continues learning new knowledge and technologies for further improvement.

4.7 Next Generation

Liam's two sons have followed their father's path and joined Liam in practicing self-sufficient farming by utilizing the expertise acquired from their education, specializing in landscape and alternative medicine. After receiving his undergraduate degree, the first son, Apakorn, received a scholarship to study in Japan from the Japanese Agricultural Exchange Council at the age of 25. This scholarship to young farmers from Thailand provided an opportunity for them to apply their knowledge upon return. Apakorn stayed in Japan for one year, during which he stayed with a local Japanese farming family as a gardener for ten months. Here, he learned about tree planting and the importance of paying attention to every single detail about the soil, weather, and daily environmental conditions. Although the farmers in Japan do not practice sufficiency economy, the knowledge and experience he acquired are now used in Ban Na Isan.

5 Lessons Learned and the Way Forward

His Majesty King Bhumibol Adulyadej's philosophy of Sufficiency Economy and the New Theory of Agriculture has allowed a number of small-scale farmers to shift their style of farming from cash cropping to farming for self-sufficiency. Liam's story from Ban Na Isan shows that a farming style focused on self-sufficiency and self-reliance can be achieved by small-scale farmers in Thailand.

Cash cropping, which is practiced by most small-scale farmers, focuses on the cultivation of only one or a few kinds of crops. Monoculture cash cropping without proper management can lead to lower crop yields due to soil degradation, topsoil runoff, and pests. Furthermore, the overuse of chemical fertilizer and pesticides to increase production harms the environment. In contrast, farming for self-sufficiency promotes balance. Rice as well as a wide variety of vegetables, herbs, trees, alternative food sources, medicines, and daily necessities can be directly sourced from one's own field and forest, which is more sustainable.

The case of Liam's family proves that the self-sufficient farming allows his family to always have food, and the surplus from their harvest can generate additional income, as demonstrated by the eighteen kinds of home-made products developed and processed by the family. By ensuring that food and necessities for personal consumption are prioritized and expenditures are minimized, the life of small-scale farmers can become more stable and less vulnerable to external shocks.

As Liam recalled, his lack of appropriate knowledge about monoculture cash cropping and his family's financial situation as well as doubtlessly following the advice of a local broker are some of the reasons why he was in debt. In contrast, the knowledge of plants and their efficacy that he learned improved the condition of Liam's family. Therefore, knowledge is a powerful tool in achieving a self-sufficient and self-reliant life. This is summarized in the 'three knows' concept employed by Wiboon know yourself, know your problems, and know the resources to understand your life. Learning centres such as the Agroforestry for Social Foundation, Mab Auang Agricultural Center, and Baan Suan Oonson and its network are important for learning, raising awareness, developing knowledge, and capacity building for many people, ranging from local small-scale farmers to government officials.

Self-sufficient farming can be perceived as an old style of farming; however, the case of Ban Na Isan shows that it is an opportunity. Liam's sons have studied landscapes and medicinal plants to further develop not only the knowledge gained from living and their father but also specialized knowledge gained from school. This reveals that the life of small-scale farmers is not an outdated form of living, and 'poor' farmers have the potential to continue this practice in the future. However, despite success stories such as Liam's, not all village households practice this style of farming. According to Liam, he initially thought that most people from Ban

Na Isan would shift their style of farming from monocropping to self-sufficient farming within the next few years. However, he is now sceptical and believes that it will require 80 years to achieve this. To change the conventional agricultural style, farmers' mindsets toward farming for self-sufficiency must first be changed, which seems quite challenging. In addition, despite the initiatives in Ban Na Isan to achieve community-level self-reliance through the establishment of a community cooperative bank, rice bank, fish bank, and community forest, only the community forest has persisted, whereas the community cooperative bank and rice bank were unable to attract attention due to government policy. Activities that require such cooperation in scale are not necessarily continued when short-term opportunities for cash acquisition are presented.

6 Conclusion

This chapter introduces the practice of farming for self-sufficiency conducted in the village of Ban Na Isan, at the east end of the Chachoengsao Province in eastern Thailand. Many small-scale farmers in rural areas in Thailand who practice monoculture cash cropping, based on government policies adapted in the 1960s, experience difficulties and are not able to improve their poor living conditions. In Ban Na Isan, most of the households migrated from Isan, an area known as the 'poor area' of Thailand. Following the Sufficiency Economy and New Theory of Agriculture advocated by His Majesty King Bhumibol Adulyadej, twelve households in Ban Na Isan shifted from monoculture cash cropping to farming for self-sufficiency. A balanced use of farmland, crop diversification, and sustainable soil and water management allowed these farmers to be self-sufficient on food and even generate cash income from the sale of excess crops. This practice has allowed rice, various vegetables, herbs, and trees, food sources, medicines, and daily necessities to be sourced directly from the field and forest in the village. This demonstrates that farming for self-sufficiency has made both the life of farmers and environment more sustainable.

In Ban Na Isan, activities at the community level have been developed from the individual level. These include the establishment of a community cooperative bank, rice bank, fish bank, and community forest. However, not all of these initiatives have been continued due to external influences. Nevertheless, such successful examples have become places of education for a variety of people, including local students and farmers. Despite being perceived as an old style of farming, self-sufficient farming, as exemplified in Ban Na Isan, is a practice that promotes a stable and satisfied life. However, not all villagers have adopted this farming style. This poses a challenge to transform the mindset of local farmers and their way of living. Self-sufficient farming can become a solution to the problem of conventional monoculture cash crop farming, particularly the risk of external price shocks and unsustainable resource management. Farming for self-sufficiency has certainly become one of the strategies for small-scale farmers to attain sustainable and stable living by taking advantage of the characteristics of the environment. The utilization of the experience and expertise of Liam's sons in this practice demonstrate its potential to lead sustainable agricultural practices for the next generation.

References

Brahmasubha, A. (2012). *BOR WORN; Home, Temple and School (HTS) Organisation: The learning organisation in the communities of Thailand* (pp. 111–123). Unpublished Ph.D. thesis, University of Hull.

Chachoengsao. (2016). *Amphoe Sanam Chai Khet*. Retrieved January 25, 2018, from http://www.chachoengsao.go.th/ccoprovince/index.php/2016-02-15-05-52-31/11/113-2016-02-23-03-05-55

Gotoknow. *Wiboon Khemchalerm: Agriculture for life and the environment*. Retrieved January 23, 2018, from https://www.gotoknow.org/posts/630041

Green Net. *Wanakaset-self reliance*. Retrieved January 24, 2018, from http://www.greennet.or.th/en/article/1023

Jitsanguan, T. (2001). *Sustainable agricultural systems for small-scale farmers in Thailand: Implications for the environment*. Food and Fertilizer Technology Center.

Kasem, S., & Thapa, G. B. (2012). Sustainable development policies and achievements in the context of agriculture sector in Thailand. *Sustainable Development, 20*, 98–114.

Leturque, H., & Wiggins, S. (2011). *Thailand's progress in agriculture: Transition and sustained productivity growth*. Overseas Development Institute.

Mongsawad, P. (2010). The phylosophy of the sufficiency economy: A contribution to the theory of development. *Asia-Pacific Development Journal, 17*(1), 123–143.

National Statistical Office (1). (2013). *2013 agricultural census whole kingdom*. National Statistical Office.

National Statistical Office (2). (2013). *The 2013 household socio-economic survey whole kingdom*. National Statistical Office.

Piboolsravut, P. (2004). Sufficient economy. *ASEAN Economic Bulletin, 21*(1), 127–134.

Pruksanubal, B. (2016). Land use transformation process in Chachoengsao Province, Thailand. *Procedia—Social and Behavioral Sciences, 222*, 772–781.

Srimongkol, K., & Marten, G. G. (1986). Traditional agriculture in Northern Thailand. In G. G. Marten (Ed.), *Traditional agriculture in Southeast Asia: A human ecology perspective* (pp. 85–102). Westview Press.

Wibulswasdi, C., Piboolsravut, P., & Pootrakool, K. (2012). *Sufficiency economy philosophy and development*. The Crown Property Bureau.

Conclusion

Miki Yoshizumi and Takaaki Miyaguchi

Abstract This concluding chapter synthesizes the key insights from the diverse case studies and conceptual frameworks presented in this volume. It emphasizes the multi-faceted nature of sustainable local food systems, highlighting the need for integrated approaches that connect environmental sustainability, economic viability, and social inclusion. Drawing on examples from Asia and Europe, the chapter underscores the importance of community-led governance, adaptive agricultural practices, and resource-based resilience in addressing climate change, market volatility, and demographic shifts. It also explores policy and research directions, advocating for locally grounded strategies, participatory decision-making, and the scalability of successful models. Ultimately, the chapter calls for a holistic and adaptive vision to promote sustainable, inclusive, and future-ready food systems worldwide.

Keywords Local food systems · Community resilience · Food governance · Sustainable agriculture · Policy integration · Adaptive capacity

The chapters in this volume, *Navigating Local Sustainability in Food, Community, and Inovation*, have explored a wide range of themes related to the sustainability, resilience, and evolution of local food systems. Through case studies, theoretical frameworks, and empirical research, the contributing authors have shed light on the challenges and opportunities that local food systems face in a variety of contexts, from rural agricultural communities to urban centres striving for food sovereignty. These discussions highlight the urgent need for interdisciplinary approaches that incorporate ecological, social, and economic dimensions to achieve long-term sustainability.

M. Yoshizumi (✉)
College of Gastronomy Management, Ritsumeikan University, Kusatsu, Japan
e-mail: miki-yo@fc.ritsumei.ac.jp

T. Miyaguchi
Faculty of Foreign Language Studies, Kansai University, Suita, Japan
e-mail: t_miyagu@kansai-u.ac.jp

M. Yoshizumi et al. (eds.), *Navigating Local Sustainability in Food, Community, and Innovation*, https://doi.org/10.1007/978-981-96-8963-7_14

1 Key Findings and Contributions

One of the central themes that emerges throughout this book is the intricate interplay between environmental sustainability, economic viability, and social inclusivity. Local food systems do not exist in isolation; they are part of a complex web of relationships involving producers, consumers, policymakers, and environmental factors. The case studies illustrate the need for an integrated approach that considers not only the production of food but also its distribution, consumption, and disposal.

Several key findings have been highlighted in the chapters:

Resilience and Adaptation: Many local food systems are under increasing pressure because of climate change, urbanisation, and changing economic conditions. Several strategies, including agroecological practices, carbon credit initiatives, and agritourism, have the potential to improve resilience. The incorporation of adaptive mechanisms such as sustainable soil management and diversified cropping systems is crucial for mitigating climate-related risks. Furthermore, the importance of local governance in promoting adaptation strategies should be emphasised.

Chapters on community-based agricultural projects demonstrate how traditional farming methods, when modified to meet modern sustainability criteria, improve resilience. For example, a case study from Thailand demonstrated how local knowledge supports sustainable food systems by highlighting the role of agroforestry in maintaining soil health and lowering reliance on external inputs. Research on the Nam O fish sauce villages in Vietnam shows that small-scale traditional food production may adapt to external shocks while maintaining cultural heritage and food security.

Community Engagement and Governance: Local governance institutions, cooperatives, and community-led initiatives are critical for strengthening food systems. Participatory initiatives that empower local actors and ensure equitable access to food resources are effective in sustaining local food networks. The involvement of multiple stakeholders fosters social cohesion and collaborative decision-making, both of which are critical to the long-term viability of local food systems. It is crucial to strengthen governance structures that allow for decentralised decision-making and participatory policies.

Several chapters emphasise the importance of farming cooperatives in promoting sustainable agricultural practices. Moreover, the role of women in local food governance is discussed, with a focus on their contribution to food security and sustainable resource management. The research on traditional food SMEs in Central Vietnam illustrates the significance of small and medium-sized enterprises (SMEs) in ensuring local food security and economic stability, particularly in times of crisis, such as the COVID-19 pandemic. The case study of Indonesia's non-rice food banks demonstrates how local institutions can improve food security by diversifying food sources and implementing community-driven initiatives, lowering dependency on imported staples. Furthermore, research on food banks in Japan emphasises their role in ensuring the right to food, particularly during economic crises, such as the COVID-19 pandemic. This study demonstrates how food banks act as critical safety nets for vulnerable populations that lack access to consistent income or food security measures.

Furthermore, the *Shinjyo Aigoukai* initiative research in Japan provides a valuable perspective on community resilience and self-sufficiency. While not directly focused on food systems, this study highlights how local communities can pool their resources to improve their safety and preparedness without relying on government aid. This model of community-led disaster resilience is extremely pertinent to local food security discussions. As food production becomes more specialised and geographically fragmented, ensuring food security at the local level requires a similar approach in which communities proactively use local resources and knowledge to build sustainable food systems that are resistant to external disruption. The *Shinjyo Aigoukai* case demonstrates how a self-sustaining, community-led approach to resource management and risk mitigation might inform the development of localised, sustainable food networks.

Yoshizumi and Miyaguchi's review chapter emphasised the broader importance of community-driven models in building local food system resilience. Their research discussed how local communities that followed sustainable food practices and governance could build more adaptive and self-sufficient food networks. The integration of food systems theory with resilience and social-ecological system frameworks suggests that communities should play an active role in developing food security strategies, ensuring that food production, distribution, and access are responsive to local needs and environmental conditions. Localised governance, knowledge-sharing platforms, and participatory decision-making can help communities reduce their vulnerability to external shocks and build a more resilient food future.

Soil fertility and organic rice farming in Japan are explored in the chapter by Thinh, which emphasises the importance of sustainable soil management for long-term food security. These data suggest that using organic fertilisers significantly increases soil microbial activity and nutrient availability, reduces dependency on chemical inputs, and promotes environmentally friendly rice cultivation. This study contributes to the broader discussion on food system sustainability by demonstrating how soil health directly influences food security and resilience.

Moreover, Tam highlights the policies and challenges of sustainable rice farming in Shiga Prefecture. This case study demonstrates how local governments play a critical role in promoting sustainable agricultural practices through direct payment schemes, environmentally friendly certification systems, and conservation initiatives, such as the Fish Cradle Rice Paddies Project. These initiatives protect biodiversity and promote agricultural sustainability by encouraging farmers to use eco-friendly farming methods. Shiga's model, which combines regulatory incentives with conservation practices, provides a valuable framework for other regions looking to promote sustainable food systems. Holger presents three German community initiatives that demonstrate how sustainable resource use may be effectively integrated into the regional economy. The preservation of alpine pastures in Bad Hindelang supports ecotourism while also upholding traditional farming practices and reinforcing the link between sustainable agriculture and economic viability. In Templin, the School

Forest Initiative emphasises the role of environmental education in fostering community engagement and increasing sustainability awareness. The Energy Cooperative in Schönau illustrates how local renewable energy initiatives create self-sufficient and resilient communities. These case studies show how important sustainable resource management and good community governance are in ensuring the sustainability of the overall food system.

2 Future Directions and Policy Implications

The insights gained from this collection of studies highlight several critical areas for future research and policymaking.

Policy Integration: To ensure sustainable local food systems, policies must connect agricultural development, environmental protection, and economic incentives.

Education and Capacity Building: Promoting sustainable food practices among consumers, producers, and policymakers is essential.

Community-Driven Models: The *Shinjyo Aigoukai* model, Yoshizumi and Miyaguchi's review, Shiga's policy initiatives, and German community initiatives emphasise the importance of locally regulated food systems in fostering resilience.

Scalability and Replication: Many successful local food initiatives discussed in this volume can be replicated and scaled in different contexts.

3 Final Thoughts

As we move forward in addressing the complexities of sustainable local food systems, no single solution exists. The key to sustainability is a holistic and adaptive approach that appreciates local knowledge, fosters innovation, and ensures inclusivity. By continuously sharing our experiences, research, and best practices, we can help promote sustainable, resilient, egalitarian, and future-ready food systems.

MIX
Papier aus verantwortungsvollen Quellen
Paper from responsible sources
FSC® C105338

If you have any concerns about our products,
you can contact us on
ProductSafety@springernature.com

In case Publisher is established outside the EU,
the EU authorized representative is:
Springer Nature Customer Service Center GmbH
Europaplatz 3, 69115 Heidelberg, Germany

Printed by Libri Plureos GmbH
in Hamburg, Germany